TRANSPORTATION ISSUES, POLICIES AND R&D

AIRCRAFT TRACKING AND FLIGHT DATA RECOVERY

BACKGROUND AND ENHANCEMENT PROPOSALS

Transportation Issues, Policies and R&D

Additional books in this series can be found on Nova's website under the Series tab.

Additional e-books in this series can be found on Nova's website under the e-book tab.

TRANSPORTATION ISSUES, POLICIES AND R&D

AIRCRAFT TRACKING AND FLIGHT DATA RECOVERY

BACKGROUND AND ENHANCEMENT PROPOSALS

JORDAN MCCOY
EDITOR

New York

Copyright © 2015 by Nova Science Publishers, Inc.

All rights reserved. No part of this book may be reproduced, stored in a retrieval system or transmitted in any form or by any means: electronic, electrostatic, magnetic, tape, mechanical photocopying, recording or otherwise without the written permission of the Publisher.

We have partnered with Copyright Clearance Center to make it easy for you to obtain permissions to reuse content from this publication. Simply navigate to this publication's page on Nova's website and locate the "Get Permission" button below the title description. This button is linked directly to the title's permission page on copyright.com. Alternatively, you can visit copyright.com and search by title, ISBN, or ISSN.

For further questions about using the service on copyright.com, please contact:
Copyright Clearance Center
Phone: +1-(978) 750-8400   Fax: +1-(978) 750-4470   E-mail: info@copyright.com.

NOTICE TO THE READER

The Publisher has taken reasonable care in the preparation of this book, but makes no expressed or implied warranty of any kind and assumes no responsibility for any errors or omissions. No liability is assumed for incidental or consequential damages in connection with or arising out of information contained in this book. The Publisher shall not be liable for any special, consequential, or exemplary damages resulting, in whole or in part, from the readers' use of, or reliance upon, this material. Any parts of this book based on government reports are so indicated and copyright is claimed for those parts to the extent applicable to compilations of such works.

Independent verification should be sought for any data, advice or recommendations contained in this book. In addition, no responsibility is assumed by the publisher for any injury and/or damage to persons or property arising from any methods, products, instructions, ideas or otherwise contained in this publication.

This publication is designed to provide accurate and authoritative information with regard to the subject matter covered herein. It is sold with the clear understanding that the Publisher is not engaged in rendering legal or any other professional services. If legal or any other expert assistance is required, the services of a competent person should be sought. FROM A DECLARATION OF PARTICIPANTS JOINTLY ADOPTED BY A COMMITTEE OF THE AMERICAN BAR ASSOCIATION AND A COMMITTEE OF PUBLISHERS.

Additional color graphics may be available in the e-book version of this book.

Library of Congress Cataloging-in-Publication Data

ISBN: 978-1-63483-930-3

Published by Nova Science Publishers, Inc. † New York

CONTENTS

Preface		vii
Chapter 1	Aviation Safety: Proposals to Enhance Aircraft Tracking and Flight Data Recovery May Aid Accident Investigation, but Challenges Remain *United States Government Accountability Office*	1
Chapter 2	Safety Recommendation A-15-1 through -8 *Christopher A. Hart*	55
Index		71

PREFACE

The crash of Air France Flight 447 (AF447) off the coast of Brazil in June 2009 and the disappearance of Malaysia Airlines Flight 370 (MH370) in the southern Indian Ocean in March 2014 highlight several challenges authorities may face in locating aircraft in distress and recovering flight recorders. In response to these aviation accidents, government accident investigators, international organizations, and industry have offered proposals that aim to enhance oceanic flight tracking and flight data recovery on a global scale. Given the implications for the U.S. commercial fleet, it is essential that the Congress understand the strengths and weaknesses of these proposals. This book describes the challenges in tracking aircraft and recovering flight data highlighted by recent commercial aviation accidents over oceanic regions; government and industry proposals to enhance aircraft tracking, and how aviation stakeholders view their strengths and weaknesses; and government and industry proposals to enhance the recovery of flight data, and how aviation stakeholders view the proposals' strengths and weaknesses.

In: Aircraft Tracking and Flight Data Recovery ISBN: 978-1-63483-930-3
Editor: Jordan McCoy © 2015 Nova Science Publishers, Inc.

Chapter 1

AVIATION SAFETY: PROPOSALS TO ENHANCE AIRCRAFT TRACKING AND FLIGHT DATA RECOVERY MAY AID ACCIDENT INVESTIGATION, BUT CHALLENGES REMAIN[*]

United States Government Accountability Office

WHY GAO DID THIS STUDY

The AF447 and MH370 disasters have raised questions about why authorities have been unable to locate passenger aircraft. In response to these aviation accidents, government accident investigators, international organizations, and industry have offered proposals that aim to enhance oceanic flight tracking and flight data recovery on a global scale. Given the implications for the U.S. commercial fleet, it is essential that the Congress understand the strengths and weaknesses of these proposals.

GAO was asked to review efforts to enhance aircraft tracking and flight data recovery. This report describes (1) the challenges in tracking aircraft and recovering flight data highlighted by recent commercial aviation accidents over oceanic regions; (2) government and industry proposals to enhance

---

[*] This is an edited, reformatted and augmented version of a United States Government Accountability Office publication, No. GAO-15-443, dated April 2015.

aircraft tracking, and how aviation stakeholders view their strengths and weaknesses; and (3) government and industry proposals to enhance the recovery of flight data, and how aviation stakeholders view the proposals' strengths and weaknesses. GAO reviewed reports by government accident investigators and others, and technology presentations by avionics manufacturers, including current cost data, which was not available in all cases. GAO also interviewed 21 aviation stakeholders, including FAA, the National Transportation Safety Board, and industry, selected based on their expertise in aviation technology and flight operations. FAA and NTSB provided technical comments on a draft of this report, which were incorporated as appropriate.

WHAT GAO FOUND

The crash of Air France Flight 447 (AF447) off the coast of Brazil in June 2009 and the disappearance of Malaysia Airlines Flight 370 (MH370) in the southern Indian Ocean in March 2014 highlight several challenges authorities may face in locating aircraft in distress and recovering flight recorders. First, oceanic surveillance is limited, and an aircraft's position may not be precisely known. For example, MH370 continued to fly for several hours outside of radar coverage after onboard communications equipment were no longer working, according to investigators. Additionally, communication and coordination between air traffic control centers in oceanic regions pose challenges. Finally, these accidents show that investigators may have difficulty locating and recovering flight recorders, which are used to determine accident causes, because of the ocean's depth and terrain. For instance, locating AF447's flight recorders took 2 years at a cost of approximately $40 million.

Proposals to enhance aircraft tracking: Following the disappearance of MH370, the international aviation community developed voluntary performance standards to establish a near-term aircraft tracking capability using existing technologies and a long-term comprehensive aircraft tracking concept of operations.

- *Near-term voluntary aircraft tracking performance standards:* An industry task force called for automatic position reporting to airlines every 15 minutes and faster reporting when triggers, such as an unusual change in altitude, are met. According to stakeholders, existing technologies can meet this standard, and many domestic

long-haul aircraft are equipped to do so, although some additional ground infrastructure may be needed. However, some airlines may face costs to equip aircraft with these technologies. In the longer term, technologies like satellite-based surveillance may provide global aircraft tracking.
- *Long-term global aeronautical distress system:* The International Civil Aviation Organization has proposed a long-term framework, which is designed to ensure an up-to-date record of aircraft progress and, in the case of emergency, the location of survivors, the aircraft, and its flight recorders. Stakeholders noted that the new framework begins to address the need for improved coordination and information sharing. One component is a tamperproof distress tracking system, which is not yet available.

Proposals to enhance flight data recovery: Low-cost actions are planned to increase the battery life of the underwater locator beacon—which emits a "ping" to help locate the flight recorders—from 30 to 90 days. In the longer term, two proposals seek to enable flight data recovery without underwater retrieval; however, neither would eliminate investigators' need to recover the wreckage itself or eliminate all search and recovery costs.

- *Automatic deployable recorders:* Designed to separate automatically before a crash and float, deployable recorders may be easier to recover. However, stakeholders are divided on equipping the commercial fleet. Some raised concerns that safety testing is needed and that equipage costs are high and potentially unnecessary given the rarity of oceanic accidents.
- *Triggered transmission of flight data:* Transmitting data automatically from the aircraft during emergencies would allow for some post-flight analysis when the flight recorders cannot be easily recovered. However, some stakeholders raised feasibility and data protection concerns.

ABBREVIATIONS

ACARS   Aircraft Communications Addressing and Reporting System
ADS-B   Automatic Dependent Surveillance-Broadcast
ADS-C   Automatic Dependent Surveillance-Contract

AF447	Air France Flight 447
ANSP	air navigation service provider
ASIAS	Aviation Safety Information Analysis and Sharing
ATC	air traffic control
BEA	Bureau d'Enquêtes et d'Analyses (French investigation authority for civil aviation accidents or incidents)
CVR	cockpit voice recorder
FAA	Federal Aviation Administration
FANS	Future Air Navigation System
FDR	flight data recorder
FOQA	Flight Operational Quality Assurance
GADSS	Global Aeronautical Distress and Safety System
HF	high frequency
IATA	International Air Transport Association
ICAO	International Civil Aviation Organization
NAS	National Airspace System
NextGen	Next Generation Air Transportation System
NTSB	National Transportation Safety Board
MH370	Malaysia Airlines Flight 370
SSMT	System Safety Management Transformation
TSO	Technical Standard Order
ULB	underwater locator beacon
UTC	Coordinated Universal Time
VHF	very high frequency

* * *

April 16, 2015

Congressional Requesters:

Although air travel is safe, aviation accidents may result in loss of life and undermine public confidence in the aviation system. The crash of Air France Flight 447 (AF447) in June 2009 and the disappearance of Malaysia Airlines Flight 370 (MH370) in March 2014 have raised questions about how—in a world of modern, interconnected technologies—authorities have been unable to locate state-of-the-art passenger aircraft. For example, more than a year after disappearing en route to Beijing, investigators have found no trace of

MH370, amid a search and recovery effort that has involved 26 nations and is projected to be the most expensive in aviation history. In oceanic regions, authorities have also faced challenges retrieving planes' flight data and cockpit voice recorders, which play an important role in helping investigators determine the causes of and circumstances surrounding aviation accidents. In the case of AF447, the effort to locate and retrieve the plane's recorders more than 12,000 feet below the ocean's surface lasted almost 2 years after the plane crashed. In response to these aviation accidents, governments, international organizations, and industry have proposed new strategies and technologies that aim in different ways to address the challenges of locating commercial aircraft and recovering flight recorders in oceanic areas.

Given the implications for the U.S. commercial fleet and Congress's role in enhancing and overseeing aviation safety, you asked us to review efforts to enhance aircraft tracking and flight data recovery. Specifically, this report describes (1) the challenges in flight tracking and flight data recovery that are highlighted by recent commercial aviation accidents over oceanic regions, (2) government and industry proposals to enhance aircraft tracking, and how aviation stakeholders view the strengths and weaknesses of these proposals, and (3) government and industry proposals to enhance the recovery of flight data, and how aviation stakeholders view the strengths and weaknesses of these proposals.

To identify the challenges in tracking commercial flights and recovering flight recorders highlighted by recent commercial aviation accidents, we reviewed government reports on the investigations of the AF447 and MH370 accidents and interviewed officials from the Federal Aviation Administration (FAA), National Transportation Safety Board (NTSB), and the International Civil Aviation Organization (ICAO) to obtain their perspectives on recent aviation accidents.[1] To determine the various government and industry proposals to enhance aircraft tracking and flight data recovery, we reviewed recommendations and reports from government investigative agencies,[2] working papers and reports from industry meetings held by international organizations,[3] technology presentations from avionics manufacturers,[4] and industry white papers.[5] We requested information from vendors and airlines regarding costs of various technology proposals. However, cost data were not available in some cases, and in other cases we were unable to obtain specific costs due to concerns about the sensitivity of proprietary information. Therefore, we provide broad estimates of potential costs where possible and discuss the types of costs that may be incurred, such as installation, data transmission, and maintenance costs. To describe the views of aviation stakeholders regarding the strengths and weaknesses of the government and industry proposals, we interviewed 21 stakeholders, including FAA and NTSB officials and representatives from two international organizations, faculty from one academic institution, two airframe manufacturers, three domestic airlines, four avionics manufacturers, two air transport communications service providers, three industry trade associations, and two satellite services companies. We identified and selected these stakeholders to include facets of the aviation industry that would potentially be affected by any new international standards or domestic regulatory changes that may be considered to enhance global flight tracking and flight data recovery. We also selected stakeholders based on their expertise with respect to aviation safety, technology, and flight operations. Their views should not be used to make generalizations about the views of all industry stakeholders, but do provide a range of perspectives on issues affecting the industry. See appendix I for a listing of the aviation stakeholders we interviewed.

We conducted this performance audit from May 2014 through April 2015 in accordance with generally accepted government auditing standards. Those standards require that we plan and perform the audit to obtain sufficient, appropriate evidence to provide a reasonable basis for our findings and conclusions based on our audit objectives. We believe that the evidence obtained provides a reasonable basis for our findings and conclusions based on our audit objectives.

BACKGROUND

On a typical day, approximately 100,000 flights around the world reach their destinations without incident, and the safety of the global air transportation system has continued to improve in recent years. The United States air transport system, in particular, is experiencing one of the safest periods in its history. The FAA, an agency of the Department of Transportation, is primarily responsible for the advancement, safety, and regulation of civil aviation, as well as overseeing the development of the air traffic control (ATC) system. The FAA's stated mission is to provide the safest, most efficient aerospace system in the world.[6] Air traffic control on a global level is coordinated by ICAO, which establishes global standards for air navigation, air traffic control, aircraft operations, personnel licensing, airport design, and other issues related to air safety.[7]

FAA collaborates with ICAO in setting standards and procedures for aircraft, personnel, airways, and aviation services domestically and throughout the world.

Air navigation service providers (ANSPs) are organizations authorized to provide air navigation services. For example, FAA's Air Traffic Organization is responsible for providing safe and efficient air navigation services in U.S. airspace. Air traffic services units provide air traffic control, flight information, and alerting services in portions of airspace called flight information regions. According to Annex 11 to the Convention on International Civil Aviation, air traffic control is provided to prevent collisions and expedite and maintain an orderly flow of air traffic, among other things. Surveillance plays an important role in air traffic control as the ability to accurately and reliably determine the location of aircraft has a direct influence on how efficiently a given airspace may be utilized. Additionally, according to ICAO, surveillance can be used as the basis for automated alert systems, as the ability to actively track aircraft enables air traffic control to be alerted, for example, when an aircraft deviates from its altitude or route. Radar is a surveillance technology that provides the air traffic controller with an on-screen view of aircraft position.[8] Air traffic control uses radar to determine the position of aircraft and the aircraft's reported altitude at a given time when traveling over land or coastlines. An aircraft's transponder automatically transmits a reply when it receives an interrogation radio signal from ground radar stations. During the cruise portion of a flight within radar coverage, the aircraft's position is reported at least every 12 seconds, depending on the rotational speed of the ground radar antenna. FAA and its aviation counterparts in other parts of the world—including Europe, Asia, and Australia—are in the process of transitioning from radar-based surveillance to a system using Automatic Dependent Surveillance-Broadcast (ADS-B), which once implemented, is expected to provide air traffic controllers and pilots with more accurate information to help keep aircraft safely separated in the sky and on runways.[9]

In areas without radar or ADS-B coverage—including oceanic airspace, remote geographic regions such as the North and South Poles, and some areas in Africa, Asia, and South America—pilots use radio[10] or satellite communications systems to report the position of their aircraft to air traffic control. According to FAA guidance for oceanic and international operations, aircraft are to report their position to the ANSP responsible for the airspace where the aircraft is operated and should do so before passing from one flight information region to another. On routes that are not defined by designated reporting points,[11] aircraft should report as soon as possible after the first 30 minutes of flight and at hourly intervals thereafter, with a maximum interval between reports of 1 hour and 20 minutes.[12] On oceanic routes, aircraft should report their position at all designated reporting points, as applicable; otherwise, flights should report their position at designated lines of latitude and longitude. Aircraft flying in oceanic airspace may be equipped with additional avionics and satellite communications capabilities, such as the Future Air Navigation System (FANS), which creates a virtual radar environment to allow air traffic control to safely place more aircraft in the same airspace.[13] FANS software, which is integrated with an aircraft's flight management system, provides a means for digital transmission of short messages between the aircraft and ATC using radio or satellite communication systems. To fly at optimum altitudes in the North Atlantic—the busiest oceanic airspace in the world—operators were required to equip with FANS by February 5, 2015. Aircraft equipped with FANS can transmit Automatic Dependent Surveillance-Contract (ADS-C) reports, which may include information on the plane's current position and intended path to air traffic control.

Position reports sent through ADS-C transmit at defined time intervals, when specific events occur such as a sudden loss of altitude, or based on a request from air traffic control.

In addition to the role of surveillance in air traffic control, FAA requires commercial airlines conducting scheduled and nonscheduled operations under part 121 of federal aviation regulations[14] to have a flight following system in place to ensure the proper monitoring of the progress of each flight from origin to destination, including intermediate stops and diversions.[15] Major airlines monitor the progress of flights from their operational centers using technologies such as the Aircraft Communications Addressing and Reporting System (ACARS), a communications system used predominantly for transmission of short text messages from the aircraft to airline operational centers via radio or satellite communication.

When an aircraft is in distress, or does not communicate as expected, air traffic control's responsibility for providing alerting services is provided in Annex 11 to the Convention on International Civil Aviation. There are 3 phases intended to notify search and rescue services to take appropriate measures, including:

- Uncertainty phase: Established when communication has not been received from the crew within the 30-minute period after a communication should have been received.

- Alert phase: Established when subsequent attempts to contact the crew or inquiries to other relevant sources have failed to reveal any information about the aircraft.
- Distress phase: Established when more widespread inquiries have failed to provide any information, and when the fuel on board is considered to be exhausted.

ICAO members establish search and rescue regions to provide communication infrastructure, distress alert routing, and operational coordination for supporting search and rescue services.[16] Within search and rescue regions, aeronautical rescue coordination centers are responsible for the search and rescue operations prompted by an aviation accident. ICAO encourages countries, where practicable, to combine their search and rescue resources into a joint rescue coordination center with responsibility for both aeronautical and maritime search and rescue.

Search and rescue authorities may be alerted to distress situations by satellite constellations operated by the International Cospas-Sarsat Program [17] that detect transmissions from an aircraft's emergency locator transmitter.[18] The emergency locator transmitter may be automatically activated by the shock typically encountered during an emergency or manually by a member of the flight crew. Satellites detect an activated transmitter and send the signals to ground stations, which determine the transmitter's position and report to search and rescue authorities. Individual countries are responsible for providing these services.

In the rare event of a disaster, after any survivors have been rescued, physical recovery of the aircraft's wreckage and the flight data recorder (FDR) and cockpit voice recorder (CVR)—commonly referred to as the black boxes—from the crash site is a priority in order to determine the cause of and circumstances surrounding the accident.[19] After recovering the recorders, each about 20 pounds and roughly the size of a shoebox, investigators from civil accident investigation authorities—such as the NTSB in the United States or the French Bureau d'Enquêtes et d'Analyses (BEA)—download and analyze data on flight conditions and the cockpit's audio environment.[20] Other potential sources of information that can help determine the cause and circumstances of an aviation accident include any communications between the flight crew and air traffic control, radar track history, data transmitted from various systems onboard the aircraft,[21] aircraft wreckage, and the crash site.

ICAO standards for flight recorders, established in Annex 6 to the Convention on International Civil Aviation, are based on aircraft weight and provide that flight data recorders and cockpit voice recorders should retain the information recorded during at least the last 25 hours and 30 minutes of operation, respectively. 22 FAA establishes the domestic regulations, policies, and guidance for the certification and airworthiness of flight recorders, as it does for other equipment and instruments on part 121 aircraft. In general, turbine powered commercial aircraft operating under part 121 must have a flight data recorder and a cockpit voice recorder, both of which undergo extensive testing to minimize the probability of damage resulting from a crash. See figure 1 for more detailed information on the FDR and CVR components.

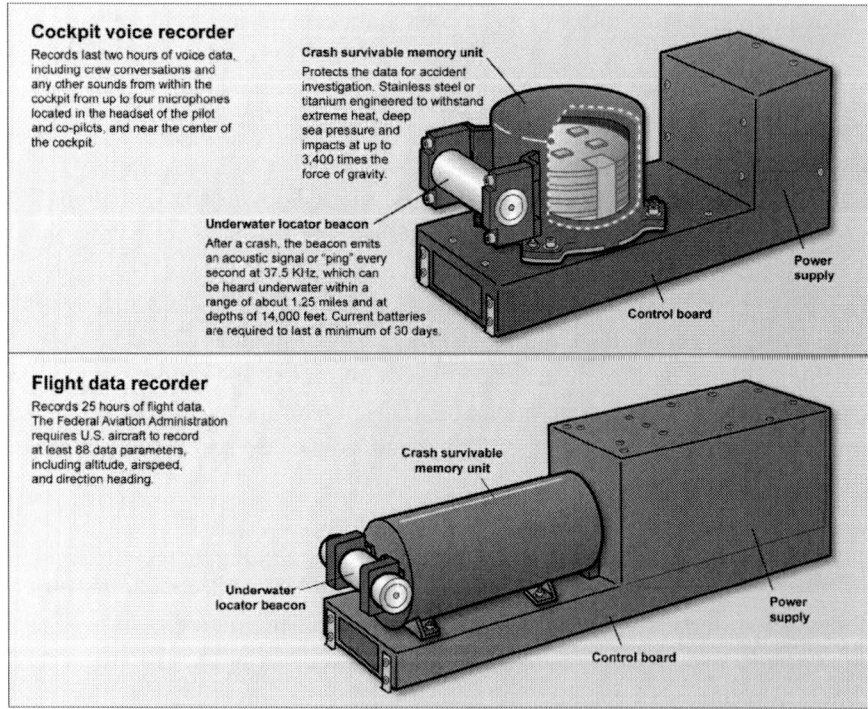

Source: Federal Aviation Administration, National Transportation Safety Board, and GAO. | GAO-15-443.

Figure 1. Flight Data Recorder and Cockpit Voice Recorder Components.

RECENT AIR DISASTERS HIGHLIGHT CHALLENGES RELATED TO SURVEILLANCE IN OCEANIC AIRSPACE, ATC COMMUNICATION, AND UNDERWATER SEARCHES

Aircraft Can Be Difficult to Locate in Oceanic Airspace because of Surveillance Limitations and the Potential for Aircraft Communications Systems to Be Disabled

Surveillance limitations in oceanic airspace and disabled aircraft communications systems may make it more difficult to determine the precise location of an aircraft in distress or an accident site. Over land, radar monitors aircraft position in real time but coverage diminishes more than 150 miles from coastlines or in remote airspace such as the polar regions. In non-radar environments, flight crews rely on procedural surveillance by periodically reporting their position to air traffic control when passing certain waypoints on their flight plan. Intervals between position reports vary, but aircraft in oceanic and remote airspace should report their position at least every 80 minutes, according to FAA guidance for oceanic operations.[23] According to one avionics manufacturer, it may take a flight crew 10 to 20 minutes to report their position using high frequency (HF) radio due to the disruptions caused by weather and atmospheric conditions over the oceans. Given that an aircraft cruises at speeds of more than 500 miles per hour (depending on altitude), an aircraft could travel more than 167 miles between 20-minute position reports, in comparison to radar's ability to determine an aircraft's location at least every 12 seconds. An aircraft reporting every 80 minutes could travel more than 600 miles between position reports. In airspace with less air traffic, such as remote oceanic regions, air traffic control does not require continuous contact with aircraft to maintain safe separation. Even more frequent position reports may not provide precise information on the aircraft's location. For example, during its scheduled flight to Paris on June 1, 2009, AF447's last regular ACARS position report was sent at 02:10 Coordinated Universal Time (UTC)[24] and 24 emergency ACARS maintenance messages were transmitted between 02:10 and 02:15, the approximate time of the plane's crash. Based on the time the last ACARS message was received, investigators established a search area of more than 17,000 square kilometers, more than 500 nautical miles from any coastline, with a radius of 40 nautical miles centered on the plane's last known location. MH370's communications systems also included

ACARS, but as discussed below, this system and other onboard systems stopped transmitting data during flight.

Additionally, air traffic control and airline operation centers may be unable to determine an aircraft's location if communications equipment onboard the plane is damaged, malfunctioning, or has been manually turned off. For example, MH370 departed from Kuala Lumpur at 16:42 UTC on March 8, 2014 on a scheduled flight to Beijing, China. According to the Australian Transport Safety Bureau, the agency leading the search for the plane, MH370's flight path includes three distinct sections:

1) an initial stage after takeoff in which the aircraft was under secondary radar, the transponder was operational, and ACARS messages were being transmitted;
2) a second stage in which onboard communications equipment were no longer working and the plane was only being tracked by military radar; and
3) a final stage in which the only available information on the flight's path comes from satellite communications log data.[25]

At 17:07, the aircraft transmitted its final automatic ACARS message, which included the weight of the fuel remaining on board. The flight crew's last radio contact with Malaysian air traffic control occurred at 17:19 and then MH370 lost contact with air traffic control during a transition between Malaysian and Vietnamese airspace at 17:22. At 18:22, Malaysian military radar tracked MH370 flying northwest along the Strait of Malacca; this was the final radar data indicating the airplane's position. After disappearing from military radar, MH370's satellite communications system exchanged seven signaling messages—also referred to as "handshakes"[26]—with the ground station, a satellite over the Indian Ocean, and the aircraft's satellite communications from 18:25 until 00:19.[27] According to the Australian Transport Safety Bureau, the final signaling message, a log-on request from the aircraft, indicates a power interruption on board that may have been caused by an exhausted fuel supply. At 01:15, MH370 did not respond to the signaling message from the ground station. Using the handshake data to determine that it continued to fly for several hours after disappearing from radar and estimates of the aircraft's range based on the fuel quantity included in the final ACARS message, investigators placed MH370's final location somewhere in Australia's search and rescue region on an arc in the southern

Indian Ocean. The current phase of the search is focused on an area of approximately 60,000 square kilometers.[28]

Communication and Coordination among ATC, Operators, and Search and Rescue Authorities in Oceanic Areas Can Be Challenging

Communication and coordination between air traffic control centers can be difficult in oceanic and remote areas. As discussed previously, responsibility for global air traffic control is divided geographically by flight information region. Over the course of an oceanic flight, an aircraft may transition frequently between flight information regions in areas in which the ability to communicate with air traffic control can be limited. With respect to AF447, the TASIL waypoint in the Atlantic Ocean is located on the boundary between the Brazilian and Senegalese flight information regions. According to the final report on the investigation of the AF447 accident, controllers from the Atlantico Area Control Center in Brazil— which had been in contact with the aircraft—and the adjacent Dakar Oceanic Area Control Center in Senegal—which never established contact with the aircraft—stated that the quality of the HF radio reception was poor the night of the accident, resulting in recurring communication problems. The report found that there were powerful cloud clusters on the route of AF447, which may have created notable turbulence. The Atlantico controller last had contact at 01:35, when the crew read off their altitude and flight plan. Over the next minute, the controller asked the crew three times for its estimated time to cross the TASIL waypoint but received no response. The flight, however, did not encounter serious problems until 02:10 and the accident occurred in the Atlantico region at approximately 02:14. The crew should have established contact with Dakar air traffic controllers at approximately 02:20 when the aircraft was due to pass the TASIL waypoint. Dakar controllers stated that they were not concerned about the absence of radio contact with AF447 given the HF problems that night and since aircraft frequently crossed all or some of the Dakar flight information region without making radio contact. The AF447 final report concluded that the radio communication problems and meteorological conditions resulted in the controllers considering the situation (i.e., no contact with the flight) as normal.

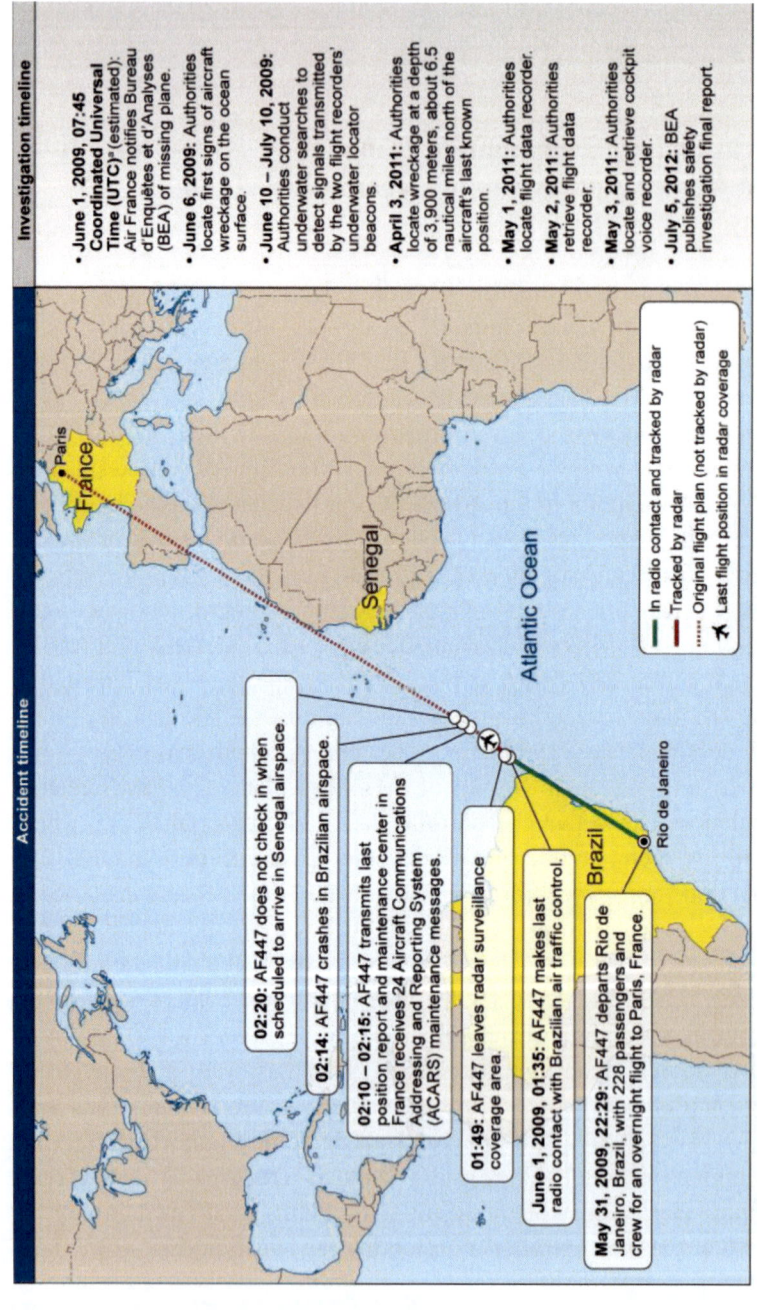

Sources: Bureau d'Enquêtes et d'Analyses and Map Resources. | GAO-15-443.
[a] Coordinated Universal Time is the 24-hour time standard common to every place in the world.

Figure 2. Timeline of the AF447 Accident and Investigation.

Furthermore, there were several communications breakdowns at critical junctures. Specifically, the report noted the lack of contact between the Atlantico controller and the flight crew before the transfer to the Dakar controller and the lack of contact between the Atlantico and Dakar controllers after AF447's projected passage of the TASIL waypoint, both of which indicated that air traffic control had not effectively monitored the aircraft. The report also noted that a timely alert was not triggered because the controller in each region failed to communicate with the other as each individual controller anticipated. See figure 2 for a timeline of the AF447 accident.

Furthermore, according to the AF447 final report, information inquiries regarding the aircraft were not coordinated, resulting in air traffic control, search and rescue, and the operators questioning each other without making a decision about what action to take. Although the last contact with AF447 occurred at approximately 01:35, it took more than 9 hours for search teams to take off from Senegal and Brazil. The first search plane arrived at the TASIL waypoint approximately 13 hours after the crash. Contrary to ICAO standards and recommended practices, Brazil and Senegal did not have a search and rescue protocol. Consequently, the Brazilian and Senegalese rescue coordination centers were not aware of each country's available resources, and the report stated that it was not possible to quickly identify one aeronautical rescue coordination center to lead the search and rescue mission. In the absence of a protocol, the report noted that the rescue coordination centers wasted considerable time gathering information and determining whether to trigger a search. The report also noted that there was a lack of coordination within the French aeronautical rescue coordination center and with its foreign counterparts in organizing the search and rescue. After being informed by Air France about a series of failure messages issued by the aircraft to its maintenance center in France, authorities from the French aeronautical rescue coordination center considered themselves not competent to intervene in a zone outside their area of responsibility. The report noted that this belief could be explained by ineffective training for search and rescue agents, particularly in terms of coordination with foreign counterparts. The French aeronautical rescue coordination center also provided key information to organizations that were not, according to the final report, competent in search and rescue; for example, one of the organizations failed to forward the last known position of the aircraft from an ACARS message.

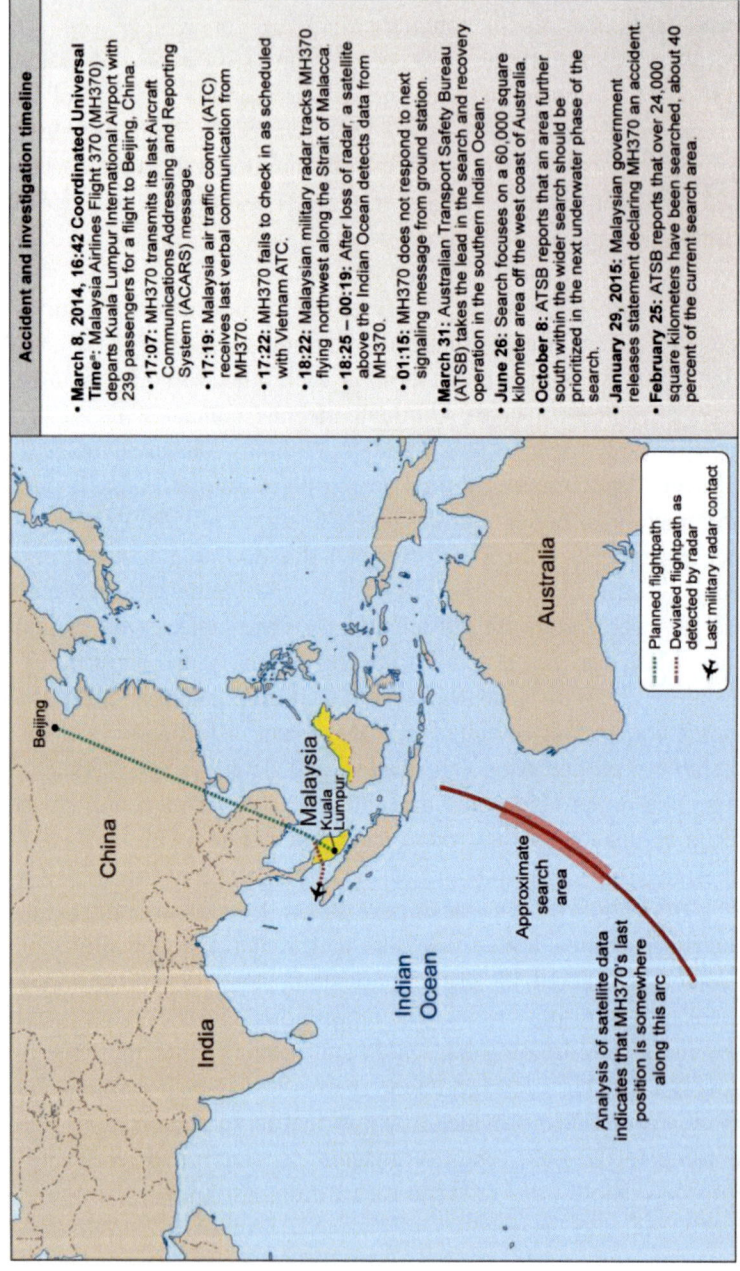

Sources: Australian Transport Safety Bureau, Ministry of Transport, Malaysia, and Map Resources. | GAO-15-443.
[a] Coordinated Universal Time is the 24-hour time standard common to every place in the world.

Figure 3. Timeline of the MH370 Accident and Investigation.

MH370 also highlights the complexities of coordinating search and rescue activities in areas with multiple flight information regions when the final location of the plane is unknown. As noted earlier, MH370 departed from Kuala Lumpur at 16:42 UTC on March 8, 2014. At 17:19, Kuala Lumpur air traffic control instructed MH370 to contact Ho Chi Minh air traffic control, and the flight crew acknowledged this request over the radio. According to the interim report issued by the Malaysia Ministry of Transport's Safety Investigation Team for MH370, another contact should have occurred at about 17:22 when MH370 passed the IGARI waypoint, but MH370 lost contact with ATC during the transition between Malaysian and Vietnamese airspace. The Ho Chi Minh City air traffic control center contacted the Kuala Lumpur air traffic control center at 17:39 to inquire about the whereabouts of MH370.[29] Thereafter, according to the report, Kuala Lumpur initiated efforts involving the Malaysia Airlines operations center and Singapore, Hong Kong, and Phnom Penh air traffic control centers to establish MH370's location, a process that lasted nearly 4 hours. The Kuala Lumpur aeronautical rescue coordination center transmitted the first distress message related to MH370 at 22:32—more than 5 hours after the last message expected from the crew—to begin search and rescue operations in the South China Sea based on the aircraft's last known position. According to the report, Malaysian search and rescue aircraft took off heading to the search areas at 03:30.

Underwater Searches for Flight Recorders Are Difficult

After an aviation accident, investigators have typically been able to recover the flight recorders in a matter of days or weeks. We reviewed data on the 16 commercial plane crashes over water that occurred globally since 2000. Additional information on each of the 16 accidents is found in appendix III. In two instances—AF447 and MH370—search and recovery efforts for the recorders exceeded 1 year. In these cases, recovery of the flight recorders was hampered because investigators did not know the precise location of the crash site. The search for AF 447 involved a 17,000 square kilometer area, and the ongoing search for the wreckage of MH370 is focused on 60,000 square kilometers in the southern Indian Ocean. Additionally, the complexities of the underwater environment may hamper efforts to retrieve recorders. While authorities located some debris from AF447 in a remote section of the Atlantic Ocean within a few days of the accident on June 1, 2009, they were unable to locate the recorders during that time. The first phase of the BEA's search for the recorders focused on the underwater locator beacons. Batteries in the current beacons are designed to allow the signal to be transmitted for at least 30 days and its range is typically limited to less than 3 nautical miles depending on water's depth, underwater topography, and surrounding conditions. If the location of the crash cannot be determined within 30 days, the time available to search for the recorders while the beacons' batteries have life is limited. The search for AF447's beacons using a towed pinger locator followed the airplane's projected trajectory in the Atlantic but was unable to detect an acoustic signal within the minimum 30-day transmission period. From July 2009 to April 2011, the BEA attempted to locate the wreckage and recorders in several search phases using sonar detection, evaluation of aircraft debris drift, and satellite-tracked buoys, each time unsuccessfully. After identifying the wreckage site at a depth of more than 12,000 feet, investigators ultimately found the flight recorders in May 2011 amid aircraft debris scattered on the seafloor. The BEA subsequently determined that the cockpit voice recorder's beacon was damaged on impact while the beacon on the flight data recorder separated and was never found. As a result of these challenges, the search for the plane's flight recorders took 2 years and cost an estimated $40 million.

NEAR-TERM PROPOSALS DO NOT FULLY ADDRESS FLIGHT TRACKING CHALLENGES, BUT LONG-TERM SOLUTIONS ARE UNDER DEVELOPMENT

In response to MH370, the international aviation community has undertaken a number of efforts to improve global aircraft tracking. In the near term, a task force formed by the International Air Transport Association (IATA) developed a set of voluntary performance standards that call for position reporting every 15 minutes with the capability to increase the reporting rate in case of emergency. Several technologies that are already on board most domestic aircraft can be used to meet this standard, although airlines would face some costs to equip if they do not already have those systems or satellite communications equipment. Over the longer term, ICAO has proposed a comprehensive new aircraft tracking framework designed to ensure that accurate information about the aircraft's location is known at all times. In addition to incorporating the industry recommendations on aircraft tracking, the new concept also proposes an autonomous distress tracking system, an alternative to underwater flight data retrieval, and new procedures to improve coordination and information sharing during emergencies.

Most Long-Haul Aircraft Are Equipped with Technologies That Can Transmit Location Information, but More Frequent Position Reporting Can Add Costs to Airlines, and Some Challenges Remain

In the aftermath of the MH370 tragedy, the international aviation community has undertaken a number of efforts to improve global aircraft tracking capabilities. For example, just weeks after the disappearance of MH370, ICAO convened a special meeting to study issues related to aircraft tracking, and international stakeholders agreed to accelerate the timeframes for a new aircraft tracking approach, according to the U.S. ambassador to ICAO. In addition, IATA, which represents the international aviation industry, formed an Aircraft Tracking Task Force (Task Force) that focused on what airlines could do to support aircraft in the near term using existing technologies. The Task Force developed a set of voluntary performance standards to establish a baseline aircraft tracking capability for all commercial passenger aircraft worldwide.[30] The key aircraft tracking performance standards proposed by the Task Force include:

- *Position reporting every 15 minutes, with capability to increase the reporting rate in response to an emergency.* This performance standard calls for regular and automatic transmission of aircraft position information.[31] The 15-minute frequency reflects the optimal balance between the benefit of knowing flight location with greater precision and the costs of transmitting data, as well as the cost of search and rescue operations, according to the Oceanic Position Tracking Improvement and Monitoring Initiative.[32] The Task Force also called for any aircraft tracking system to have the capability to report more frequently when certain circumstances are met, such as unusual change in the trajectory, vertical speed, or altitude of the aircraft.[33] The purpose of the increased position reporting rate is to narrow the search zone for an aircraft in distress.
- *Position reports should include latitude, longitude, altitude, and time information.* This performance standard calls for position reporting in four dimensions. Latitude and longitude provide the aircraft's location

on a map, while altitude and time provide other data points to pinpoint the precise position of the aircraft at any stage of the flight.
- *Communications protocols between the airline and air traffic service provider to facilitate coordination in case of an emergency situation.* The Task Force recognized that there is a need both to amend existing procedures and to develop new or improved communications protocols between airlines and air traffic service providers. The purpose of this performance standard is to establish communication procedures and protocols to better respond to instances of missing position reports or other unexplainable developments.

In order to achieve this baseline aircraft tracking standard, the Task Force recommended that aircraft operators evaluate their capabilities, implement measures to meet the performance standards within 12 months, and exchange best practices. According to the Task Force, these standards and recommendations are designed to improve the collective ability of the airline industry to identify and track aircraft globally. The Task Force also recognized that near-term procedures are just first steps in a longer-term, integrated concept of operations for aircraft tracking during all phases of flight. This concept of operations is discussed further below.

Several technologies could be used to meet the recommended aircraft tracking performance standards, according to the Task Force and aviation stakeholders we interviewed. For instance, ACARS- and FANS-equipped aircraft can be configured to report aircraft position information, even though ACARS is not specifically designed for that function and FANS is designed to report to air traffic control, not to airlines. According to the Task Force, ACARS uses information derived from the aircraft's flight management system to report the aircraft's position, and ACARS is configurable for enhanced reporting triggered by unanticipated altitude changes or flight levels below a predetermined altitude.[34] For aircraft that are equipped with FANS, the airline ground systems can be configured to access information about the position of the aircraft, using Automatic Dependent Surveillance-Contract (ADS-C)—an application that allows the airline or air traffic control to establish a contract with the FANS system onboard the aircraft to deliver four dimensional position and other data at single, periodic, or event-based intervals. In addition, other benefits of FANS include reduced separation between aircraft, more direct routings leading to reduced fuel consumption, and improved communication clarity between the pilot and air traffic control. Other commercially available systems, including FLYHT Aerospace

Solution's Automated Flight Information Reporting System, would also meet the proposed performance standards by providing operators with precise information about the aircraft's position in real-time, according to the manufacturer.[35] Passenger Wi-Fi systems, which utilize satellite connectivity, could also be used to facilitate aircraft tracking, according to representatives from one domestic airline.

The level of equipage for these various technologies differs across the U.S. and global fleet. According to one of the major air transport communications service providers, almost all commercial passenger jet aircraft operators in the U.S. install and use ACARS, including nearly all regional airlines. 36 Three major domestic passenger airlines that we spoke to also confirmed that their entire fleets are equipped with ACARS. Generally, airlines based outside the U.S. use ACARS, except some low-cost airlines that have avoided the cost of installing ACARS avionics and use only very high-frequency (VHF) voice radio or other solutions, according to one air transport communications service provider.

According to some aviation stakeholders we spoke to—including FAA, two domestic airlines, and one of the major air transport communications service providers—fewer airlines have equipped with FANS because it is only beneficial to the airlines when flying in certain, higher density oceanic airspace. FAA officials we spoke to estimated that approximately 70—80 percent of the aircraft operating in the busy North Atlantic airspace are currently FANS-equipped because it is required to access the optimal routes. According to one air transport communications service provider, FANS equipage on short-haul aircraft is very low, but is expected to increase because aircraft will need FANS avionics to be able to communicate with certain components of FAA's Next Generation Air Transportation System (NextGen) in the future.[37] Airlines that wish to take advantage of the optimal flight paths between North America and Europe will need to be FANS-equipped by 2015; therefore, FANS equipage is expected to increase in the future, according to FAA.[38] Representatives from one large domestic airline said they are installing FANS on all of their aircraft used for international operations largely because of its operational and safety benefits. Another airline that we spoke to is also planning to equip its aircraft that operate from the U.S. West Coast to Hawaii because of the operational efficiencies expected by using FANS. Finally, two major airframe manufacturers told us that all new aircraft typically come equipped with the latest communications, navigation, and surveillance avionics, including ACARS and FANS, but the operator chooses to enable the system depending on where the aircraft is used.

Despite their benefits, the technologies that could be used to achieve the Task Force's baseline aircraft tracking standard in the near term do not address all the challenges associated with locating flights. For example, according to one major airframe manufacturer, neither ACARS nor FANS is tamper-proof, which means that a knowledgeable individual could disable both systems and the aircraft's transponder. Should those systems and the transponder be turned off, the aircraft would be incapable of sending position data. Aviation stakeholders told us that there are legitimate engineering and operating reasons for the flight crew to have total control over all on-board systems. According to one major airframe manufacturer, aircraft are designed with the assumption that the pilot and flight crew are trusted and should have complete control over the aircraft. In certain situations, air traffic control may ask a pilot to turn the transponder off and back on to identify an aircraft. Other stakeholders we spoke to said that the pilot's ability to turn off any system on board the aircraft is based first and foremost on safety considerations.

Nevertheless, at least two major international airlines have called for a tamper-proof aircraft tracking solution.

Should they choose to adopt the technologies described above, airlines that currently do not meet the Task Force aircraft tracking performance standards would face some costs. Estimates of those costs across the fleet are difficult to determine with any precision because data on the level of aircraft equipage were not consistently available and the contracts between the airlines, avionics manufacturers, and air transport communications service providers to provide such services are proprietary. Costs to equip with ACARS using VHF radio could be up to $100,000 per aircraft. Additionally, ACARS using satellite communications would cost another $60,000 to $150,000 per aircraft for Iridium or Inmarsat equipment, according to one air transport communications service provider. However, according to the airframe manufacturers we spoke with, most long-haul aircraft that fly in oceanic and remote regions are already equipped with those units. For aircraft without FANS, there would be an additional cost of up to $250,000 for a new FANS-capable flight management system, according to one air transport communications service provider.[39] Costs for FLYHT Aerospace Solution's Automated Flight Information Reporting System, a commercial system that could, among other things, provide aircraft position data, were approximately $70,000 per system, including installation labor per aircraft but not the cost of data transmission, according to company representatives.[40]

In order to more frequently report position information using ACARS or FANS, airlines may have to pay for increased data transmission, but we were unable to determine the extent of these costs to industry. According to one air transport communications service provider, ACARS data transmission costs per month can range from $500 per short-haul aircraft using VHF radio systems, to approximately $1,000 per aircraft for long-haul aircraft using satellite communications over oceans. According to one domestic airline, airlines pay for ACARS messages through plans similar to cellular text messaging, and therefore, it is not clear whether more frequent position reports would be covered under existing plans, or would require new plans at a higher cost.

To help mitigate these costs and enhance tracking capabilities in the near term, aviation stakeholders have offered a number of proposals to enhance flight tracking. One proposal offered by the satellite communications provider Inmarsat would provide four free position reports per hour using FANS ADS-C capability. To take advantage of this proposal, aircraft would need to be equipped with FANS and Inmarsat satellite communications. A separate proposal from SITA, a major provider of ACARS data, aims to provide ADS-C reports to airlines. According to SITA, the company's proposal may help improve coordination between the airline and the air navigation service provider, especially if there is an unexpected event onboard the aircraft.[41] Rockwell Collins, the other major providers of ACARS data, unveiled a flight tracking service in March 2015 that utilizes several data sources, including ADS-B, ADS-C, and ACARS.[42] Over the longer term, other aircraft surveillance systems may become available that build on FAA's transition to NextGen.[43] Aireon, a joint venture between four air navigation service providers—Nav Canada, ENAV (Italy) and the Irish Aviation Authority, and Naviair (Denmark)—as well as the satellite service provider Iridium, aims to use ADS-B technology on satellites to provide a global aircraft surveillance system.[44] According to Aireon representatives, its space-based ADS-B system is scheduled to be fully deployed in 2017, although the system would not be operational until after a test and validation phase is completed, which is currently planned for early 2018.[45] The real-time surveillance costs provided by this system are being discussed with individual air navigation service providers at this time. Aviation stakeholders we spoke to recognize the potential of this spaced-based surveillance system in terms of enhancing aircraft tracking in oceanic and remote regions. Aireon has also announced a free service to be offered using the space-based ADS-B system—Aircraft Locating and Emergency Response Tracking (ALERT)—that would provide the last known or current location of any aircraft equipped with ADS-B technology to search and rescue teams in emergency situations.[46]

ICAO's Long-Term Framework for Aircraft Tracking during All Flight Phases Proposes Tamper-Proof Autonomous Distress Tracking, but Poses Several Implementation Challenges

In parallel with the Aircraft Tracking Task Force, an ICAO-led Ad-Hoc Working Group on Flight Tracking developed a long-term framework— called the Global Aeronautical Distress and Safety System (GADSS)—to ensure that accurate information about the aircraft's location is known during the sequence of events before and after an accident. Both industry and ICAO worked to harmonize their proposals, according to stakeholders involved in the process, and at ICAO's High-Level Safety Conference in February 2015, delegates from over 120 nations endorsed the GADSS framework for aircraft tracking. This framework is designed to maintain an up-to-date record of aircraft progress and, in the case of a forced landing, the location of survivors, the aircraft, and the flight recorders.[47] Conceptually, the GADSS framework incorporates the Task Force recommendations on tracking aircraft, but goes further, as described below.

The GADSS consists of four key system components:

- *Aircraft tracking system:* This tracking system incorporates the Task Force's near-term recommendations to enhance aircraft tracking, as described above, and specifies that when an abnormal event is detected, the position reporting rate of the aircraft tracking system increases to around a 1-minute interval, an increase that translates to knowing the aircraft's position within 6 nautical miles; such reporting can be achieved with the systems discussed previously.[48]
- *Autonomous distress tracking system:* The GADSS framework goes further than the Task Force standards by calling for an autonomous distress tracking system. According to the Ad-Hoc Working Group, an autonomous distress tracking system operates independently from the regular aircraft tracking system and may be automatically or manually activated at any time. This system could be automatically triggered by unusual attitude, speed or acceleration, failure of the regular aircraft tracking system or surveillance avionics, or a complete loss of engine power.[49] In addition, the system would operate independently of aircraft power or other systems, and be tamper-proof.
- *Automatic deployable flight recorder:* The GADSS proposal currently calls for an automatically deployed flight recorder. This device is

designed to automatically separate from the aircraft in the event of an accident. At the February 2015 High-Level Safety Conference, ICAO proposed the use of deployable recorders or an alternative for data retrieval. Additional information about deployable flight recorders is provided later in this report.
- *Procedures and information management:* The final component of the GADSS aircraft tracking framework recognizes that the effectiveness of any search and rescue service is only as good as the weakest link in the chain of people, procedures, systems, and information. Therefore, in addition to the technology, the GADSS identifies key areas of improvement, such as existing procedures, improved coordination and information sharing, and enhanced training of personnel in reacting to emergency circumstances.

The GADSS framework calls for an autonomous distress tracking system. According to one airframe manufacturer we spoke with, the technology to provide an autonomous distress tracking system does not exist, although a second-generation emergency locator transmitter, which is under development, could perhaps achieve this target. When activated after a crash, emergency locator transmitters send out a distress signal, which if detected by satellites, can aid in locating aircraft in distress. A second-generation emergency locator transmitter would be able to detect an emergency situation and activate automatically in flight, thereby possibly starting the transmission of position information before the crash impact, according to ICAO.

However, this technology is still in development and implementation is likely years away as the RTCA (formerly known as the Radio Technical Commission for Aeronautics) and the European Organization for Civil Aviation Equipment have been working in parallel to address the technical specifications required to design and manufacture second-generation emergency locator transmitters. These groups are also developing specifications for criteria that would trigger activation of these devices. Currently, emergency locator transmitters are not required of scheduled commercial flights due to the statutory exception, although ICAO standards call for emergency locator transmitter equipage for such aircraft and the European Aviation Safety Agency requires that commercial passenger aircraft operating in Europe be equipped with such a transmitter.

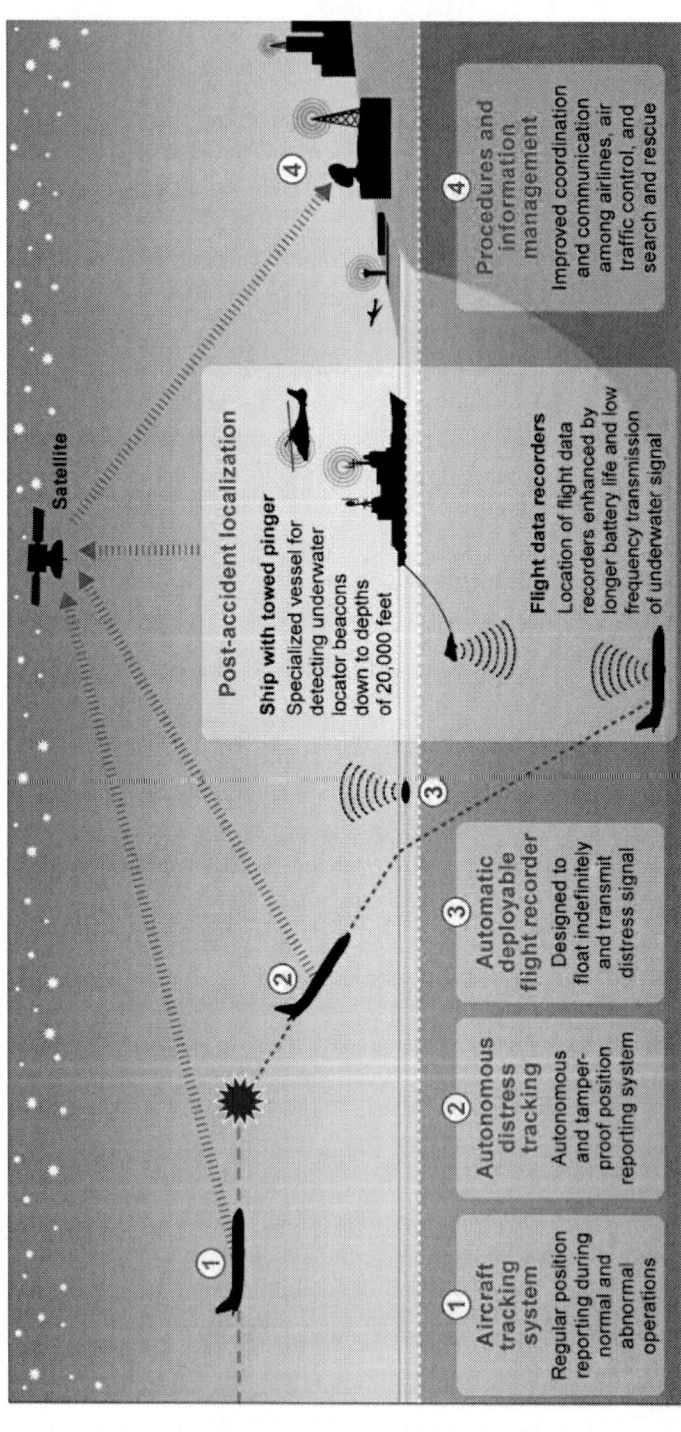

Sources: Federal Aviation Administration, International Civil Aviation Organization, National Transportation Safety Board, and GAO. | GAO-15-443.

Figure 4. Aviation Accident Scenario Based on ICAO's Global Aeronautical Distress and Safety System Framework.

U.S. aviation stakeholders expressed concern with emergency locator transmitters. For example, one airframe manufacturer and one industry association said that the current generation of transmitters has been proven not to be effective in locating aircraft wreckage to date, because the transmitters tend to get separated from their antenna upon impact and do not transmit underwater.

Several stakeholders cited other concerns related to using emergency locator transmitters as the autonomous distress tracking system. FAA officials and one airframe manufacturer told us that adding an emergency locator transmitter to an aircraft can introduce risk. For example, in 2013, an Ethiopian Airlines 787 Dreamliner was damaged by a fire caused by the lithium battery located inside the transmitter. The power source for any autonomous distress tracking system would need to be engineered so as not to introduce new safety concerns, according to one airframe manufacturer. In addition, representatives from one trade association and one airframe manufacturer cited concerns about adding equipment on an airplane that the pilot cannot turn off if need be.

In addition, stakeholders were concerned about a second-generation transmitter activating in non-emergencies. Airframe manufacturer representatives told us that it is critical that event triggers be accurate given the resource mobilization costs for a real oceanic emergency. Finally, according to aviation stakeholders we spoke with, the cost of a system that could meet the GADSS proposed autonomous distress tracking component is unknown because such a system has not been developed.

ICAO recognizes that implementing the GADSS will represent a large financial cost to the industry. The ongoing operational cost of aircraft tracking will depend on the individual solutions adopted by the airlines, according to ICAO.

In addition, more work is needed to more formally develop the specifications and performance-based standards for certain components of this new aircraft tracking concept of operations; until that work is completed, the actual cost to implement the GADSS is difficult to determine. Moreover, there may be other operational benefits that airlines can achieve by implementing aircraft tracking solutions, according to ICAO.

The components of the system would need to be more formally developed and it would likely have to pass a cost-benefit analysis performed by FAA to be adopted in the United States.[50]

Until the technologies and their associated costs to meet the GADSS are determined, it is unlikely that FAA could conduct an analysis to evaluate the costs and benefits of this long-term aircraft tracking framework.

The GADSS concept of operations also recognizes that technology is not the only solution to improving global aircraft tracking. The data supplied by the aircraft tracking, autonomous distress tracking, and the automatic deployable flight recorder components of the system must be effectively shared among all stakeholders to ensure the effective operation of the new framework.

The GADSS calls for System Wide Information Management to ensure that stakeholders share the same information in emergencies.[51] For example, when an emergency is detected, aircraft tracking information would be broadcast to the relevant stakeholders and available worldwide, but subject to agreed-upon and implemented rules. The GADSS also begins to establish the procedures that are necessary to facilitate communication during an emergency. For instance, the procedure for declaring the emergency phase is to be initiated when: (1) the regular 15-minute tracking information is not received, (2) the rate of reporting is increased, or (3) communications cannot be established within 15 minutes of the first attempt by the air navigation service provider.

As noted above, there were several communications breakdowns at critical junctures of the AF447 disaster, including a delay in triggering an alert because controllers in the Atlantico and Dakar flight information regions failed to communicate with the other as each individual controller anticipated. Furthermore, information inquiries between air traffic control, search and rescue, and the operator regarding AF447 were not coordinated, resulting in a delay of more than 9 hours to initiate search and rescue. Similarly, communication and coordination challenges among the multiple flight information regions during the early moments of the disappearance of MH370 ultimately led to delays initiating search and rescue. According to the Ad-Hoc Working Group, additional assessment of the shortcomings in coordination and information sharing between air navigation service providers and search and rescue authorities is needed. Moreover, the GADSS also calls for the development of guidance material and training on emergency situations for air navigation service providers. These efforts outlined in the GADSS are essential; however, the steps to strengthen the people, procedures, systems and information sharing must be carried out by individual countries.

IMPROVEMENTS TO FLIGHT DATA RECOVERY ARE STARTING WITH EXTENDED BEACONS, WHILE OTHER PROPOSALS MAY IMPOSE SIGNIFICANT COSTS, ACCORDING TO STAKEHOLDERS

In response to recent aviation accidents, government, international organizations, and industry have been developing proposals to enhance flight recorder recovery in oceanic regions. In the near term, manufacturers are scheduled to begin adding extended batteries on the underwater locator beacons (ULBs) attached to flight recorders and have the option to add a second low-frequency device. Longer-term proposals to equip commercial aircraft with automatic deployable flight recorders and the capability to stream up to all FDR data are in various stages of development. Some have also called for enhanced cockpit recorders to aid accident investigation by recording additional audio and adding images. While each technology or proposal is intended to improve data recovery and accident investigations, industry has raised various concerns for the commercial fleet regarding the need for such changes and the costs associated with them.

Actions under Way to Extend Battery Life and Introduce Low Frequency Devices Are Intended to Enhance Flight Recorder Recovery in the Near Term

To help address challenges in locating flight recorders and aircraft wreckage in oceanic areas, FAA and other international aviation authorities have taken actions recommended by the French BEA investigating the AF447 accident to enhance ULBs. Additionally, the NTSB has also recently issued a related recommendation. First, to help address the challenges posed by the limited time that search and rescue authorities have to detect the ULB signal, the BEA recommended extending the ULB battery life from 30 to 90 days. In February 2012, the FAA issued a Technical Standard Order (TSO) for a 90-day battery ULB effective March 1, 2015, at which point the previous TSO for a 30-day battery would be revoked.[52] The FAA issued a TSO authorization to one manufacturer in December 2014 and to another in February 2015. However, on March 10, 2015, the FAA delayed the effective date of the TSO until December 1, 2015, to provide a major aircraft manufacturer additional time for testing and analysis of the device's installation.[53] Underwater locator beacons manufactured on or after the December 1, 2015, effective date must meet the new requirements. The FAA is not requiring U.S. airlines to retrofit aircraft with a 90-day battery immediately. Instead, airlines are expected to replace 30-day ULB batteries through attrition on their normal replacement schedule of approximately every 6 years so as not to introduce additional costs associated with taking aircraft out of service outside of regular schedules. We estimate that all U.S. domestic aircraft should have a 90-day battery ULB installed by the end of 2021. Related international standards will become effective by 2018.[54]

Second, to help search and rescue authorities locate the aircraft wreckage underwater, the BEA recommended adding a mandatory second underwater locating device with a greater transmission range. The FAA issued a TSO, effective June 26, 2012, that allows manufacturers to add a second optional device directly attached to the airframe. This second device would emit a low-frequency acoustic signal with a range of approximately 5 miles, which is about four times the range of the ULBs attached to the flight recorders. This additional device is optional for airlines and manufacturers. In January 2015, NTSB issued a recommendation for a low-frequency device attached to the airframe that will function for at least 90 days and that can be detected by equipment available on military search and rescue, and salvage assets commonly used to search for and recover wreckage.[55]

Stakeholders we spoke with, including airframe manufacturers, a trade association, and an avionics manufacturer, generally agreed that extending the ULB battery life made sense and could improve flight data recovery for oceanic accidents at a low cost to the airlines; however, there was no consensus on the need for a second underwater locating device. One airframe manufacturer told us that it was taking steps to prepare for attaching a low-frequency device to the airframe, whereas another told us that this might not be necessary if an aircraft tracking solution could meet the same performance of providing an aircraft's last known location within 6 nautical miles. However, ocean currents can move aircraft wreckage from its initial point of impact, and therefore, it is unclear whether a tracking solution would provide the same function as a second device.

While both the 90-day battery ULB and the low-frequency underwater locating device are intended to help locate the flight recorders and aircraft wreckage in oceanic regions and potentially mitigate some costs with an underwater sonar search, they do not address all the potential challenges of flight recorder retrieval in remote oceanic regions. For example, investigators must still know the general location of impact. Otherwise, the search area would be too large, hampering the location and retrieval of the flight data for the accident investigation despite these longer-life and longer-range devices. The flight tracking proposals discussed previously address this concern to a certain extent. In addition, underwater conditions—including depth, topography, and surrounding conditions—can still affect ULB performance. Range can also be reduced if the device is covered or blocked by aircraft wreckage. Finally, locating the signal is only one step. Even if investigators detect the signal, retrieving the recorders may be difficult if located deep underwater or in difficult terrain.

ICAO and NTSB Have Proposed Flight Data Recovery without Underwater Retrieval, Though Some Stakeholders Identified Concerns for the Commercial Fleet

In addition to enhancing the ULBs to help locate the recorders underwater, governments, international organizations, and industry have been looking at additional changes to improve flight data recovery in oceanic regions, one prescribing a specific technology and another using a performance-based approach. ICAO included automatic deployable flight recorders in its long-term GADSS framework. NTSB issued a safety recommendation calling for a means of recovering flight data without underwater retrieval, which would allow for either a deployable recorder or triggered transmission of mandatory flight data during emergencies. While these technologies are designed to improve flight data recovery, some aviation stakeholders had concerns with installing either of these technologies on the commercial fleet.

Automatic Deployable Flight Recorders

As discussed above, ICAO's longer-term GADSS framework calls for automatic deployable flight recorders in order to provide faster and easier recovery of flight recorder data, especially in oceanic regions. Deployable recorders, which have been used for decades on U.S. military aircraft, including military versions of commercial aircraft, and helicopters, combine an FDR/CVR with an emergency locator transmitter in one crash-survivable unit.[56] The deployable recorder is designed to separate automatically from an external section of the tail or leading edge of the aircraft when sensors detect an imminent crash. After separation, the deployable recorder is designed to avoid the crash impact zone and, when over oceanic regions, to float indefinitely. These recorders are designed to deploy and emit alert messages even if the aircraft loses power. The embedded emergency locator transmitter would send Cospas-Sarsat satellites an alert every 50 seconds, including the aircraft tail number, country of origin, location of aircraft at separation, and the recorder's current location.

Several aviation stakeholders—including airframe manufacturers, trade associations, and U.S. domestic airlines—are divided in their support for implementing deployable recorders on the commercial fleet. Boeing and Airbus, for example, have publicly taken different positions. At the NTSB's forum on flight recorder technology in October 2014, Airbus representatives announced the company's tentative plans to install deployable recorders on its future A350 and A380 long-haul fleets as a second complementary system to a fixed combination FDR/CVR, though specific timing is unknown. However, at the same forum, Boeing representatives stated that the company had no plans to add deployable recorders to its fleet and that the risk for unintended consequences needed to be studied further.

Deployable recorders would offer several potential benefits, according to stakeholders, including the following:

- **Faster and easier flight data location and retrieval:** Since a properly deployed and undamaged deployable recorder floats indefinitely on the ocean surface and transmits an alert message directly to Cospas-Sarsat for up to 150 hours, it may be easier for searchers to locate and retrieve than a fixed recorder located on the seabed. In the case of AF447, if equipped with a deployable recorder that operated correctly, the device could have alerted searchers to its location, and once found on the ocean's surface, investigators would have recovered the flight data more quickly than the 2 years it took to locate and retrieve the fixed recorders from the ocean floor. Similarly, if MH370 had a deployable recorder which operated correctly in the event of a crash, investigators would generally know the location at impact and could have recovered the flight data from the ocean's surface.

- **Updated location information:** The alert messages transmitted from the recorder after deployment could help investigators pinpoint the crash site. Additionally, satellites can track the device's drift pattern from these messages for up to 150 hours, giving investigators information on ocean currents that may move survivors and debris away from the initial crash site.

- **No recurring service fees:** Data transmission of the emergency locator transmitter signals from a deployable recorder would be free because Cospas-Sarsat has no service fees.

However, some stakeholders highlighted a number of concerns with introducing deployable recorders into the commercial fleet:

- **Safety:** A range of stakeholders that we spoke to, including an airframe manufacturer, avionics manufacturers, and faculty from an academic institution, identified potential safety risks to the aircraft, passengers, maintenance technicians, and others on the ground from inadvertent deployment. According to one manufacturer, even if the industry had met its standard failure rate of less than one incident per 10 million flight hours for civil airborne systems and equipment, there would have been an estimated five incidents involving deployable recorders with the 54.9 million total commercial fleet hours in 2013. These types of incidents could potentially cause damage to both people and property. An airframe manufacturer and an avionics manufacturer told us that the system would need to be designed to ensure the safety of those on the ground and in the air, especially given the infrequency of aviation accidents in which a deployable recorder would be useful.
- **Infrequency of accidents and success of fixed recorder recovery:** One avionics manufacturer and two U.S. domestic airlines that we spoke with questioned the need for deployable recorders given the safety concerns discussed above and the infrequency of aviation accidents. Additionally, when accidents do occur, investigators typically locate and recover the flight data even for accidents occurring over water.[57] Investigators located and recovered the flight recorders in 15 of the 16 commercial accidents, approximately 94 percent of cases that occurred over water since 2000.[58]
- **Mixed recovery record in military aircraft:** According to one stakeholder, flight data recovery rates are actually better on fixed FDRs compared to deployable recorders based on experience with one military aircraft model. For instance, according to industry data of a certain military aircraft that we reviewed from 2004 through 2014, there was a 100 percent flight data recovery for fixed recorders compared to 75 percent with deployable recorders. The causes of

those failures included instances in which the recorder did not deploy, was not located, or did not have data on the memory card.

- **Does not mitigate the need to recover the aircraft wreckage and fuselage, or human remains:** While the flight recorders are an important part of the accident investigation, investigators still want to recover the aircraft wreckage and fuselage to help determine the cause of the accident. Therefore, there would still be costs for an underwater search and recovery even if investigators had the deployable recorder.
- **Hardware costs:** Deployable recorders would require adding more equipment on the commercial fleet and, according to the FAA and two avionics manufacturers, would result in additional costs to airlines. The estimated cost is $50,000—$60,000 per unit, according to an avionics manufacturer. We found that the total cost of equipage could be as much as $29 million, and any additional costs associated with certifying each aircraft type model, if this cost per unit were incurred for the current total U.S. long-haul transoceanic fleet.[59] However, for several reasons it is difficult to extrapolate the per-unit cost estimate to obtain a cost estimate for the total fleet. First, costs could vary based on the type of aircraft, the regulatory environment, and other engineering factors, so when widely deployed these factors may affect the costs for varied contexts. Additionally, if a technology becomes mandated or, even if voluntary, becomes widely adopted, unit costs might decline due to the efficiencies of mass production and also possibly due to a greater number of providers entering the market.

Given the safety concerns, costs of equipage, and the uncertain benefits associated with deployable recorders, certain stakeholders, including one airframe manufacturer, one trade association, and one avionics manufacturer, suggested additional study is needed on the use of deployable recorders in commercial aircraft. Additionally, several stakeholders raised concerns about prescribing a specific technology—in this case automatic deployable flight recorders—as opposed to a performance-based approach when implementing or enhancing flight data recovery capabilities, which is the preferred approach of both FAA and the airline industry. Several stakeholders told us that industry needs the flexibility of performance standards over prescriptive solutions due to the diversity of the fleet and to allow for technological advances. FAA officials described their position on deployable recorders as fluid and told us that they did not have plans to mandate a deployable recorder, but also would not prevent an operator from installing one provided the operator adequately demonstrated an acceptable level of safety and performance.[60] In the GADSS framework, ICAO identifies the need to develop performance-based standards for deployable recorders. At the February 2015 Second High-Level Safety Conference, ICAO presented a prescriptive baseline recommendation for deployable recorders and a performance-based alternative, though it is unclear what that alternative would be at this time. At the meeting, member states agreed with a performance-based approach to data retrieval.

Triggered Transmission of Flight Data

In January 2015, NTSB issued a safety recommendation that all new commercial aircraft used in extended overwater operations be equipped with a means of recovering mandatory flight data that does not require underwater retrieval, which builds on earlier efforts from the BEA in response to AF447. While automatic deployable flight recorders as discussed above could meet the recommendation, this performance-based approach also allows for a solution that involves triggering a transmission of mandatory flight data from the aircraft to the ground during emergencies. NTSB's approach, if adopted by FAA, allows airlines to choose a solution that fits best with their operations. The concept of triggering the transmission of flight data to the ground consists of using flight parameters to detect whether an emergency situation is forthcoming, and if an emergency arises, transmitting data automatically from the aircraft until either the emergency situation ends or the aircraft crashes.

Industry stakeholders cited various potential benefits of triggered transmissions of flight data.

- **Provides data when physical FDR cannot be recovered:** Streaming FDR data allows for post-flight analysis in instances where the physical FDR or its data cannot be easily recovered, including cases where a deployable FDR may not be recovered.
- **Faster flight data retrieval:** Triggered transmission provides required flight data without the need to recover the physical FDR, and therefore, investigators would have data available to them more quickly than if they had to initiate a potentially lengthy search for the physical FDR or deployed FDR. Quicker access to flight recorder data after an incident, especially in oceanic regions, could potentially allow investigators to determine the accident's cause more quickly. In the case of AF447, if the FDR data were streamed from the aircraft before impact, investigators could have had access to the FDR data more

quickly, which would have potentially avoided some of the lengthy search and recovery effort.

- **Provides location information for aircraft in the event of an accident:** In addition to flight data, the transmission of position information could be triggered at a rate that should allow investigators to identify a narrower search area than would otherwise be known through position reports sent at 15-minute intervals. This could minimize the need for emergency locator transmitters, which as discussed previously, several stakeholders told us have been unreliable in providing crash location information.

However, stakeholders we interviewed raised several concerns about implementing data streaming into the commercial fleet.

- **Bandwidth limitations:** Stakeholders—including the FAA, an airframe manufacturer, faculty from an academic institution, a U.S. domestic airline, and two trade associations—told us that while it would likely not be possible to stream FDR data continuously from every aircraft in flight with current satellite capabilities, it could be possible to stream FDR data in a limited number of emergencies. However, a representative from an avionics manufacturer that sells a flight data streaming product told us that streaming voice data was not technically possible at the current time due to limited satellite bandwidth. Therefore, investigators would still need to retrieve the physical CVR to obtain audio information; such retrieval would limit some of the potential benefits of triggered transmission.
- **Technical challenges:** Two avionics manufacturers and an air transport data communications provider also stated that it could be technologically difficult for an aircraft in distress with its satellite antenna not pointing in a fixed position prior to impact to transmit data to satellites. However, there is at least one commercial product—FLYHT's Automated Flight Information Reporting System—capable of streaming up to all FDR data and position information in near real-time via satellites when triggered by an onboard emergency, and according to this manufacturer, a test flight showed FDR data transmission even when the plane was in an unusual position during flight.
- **Data privacy and security:** Flight recorder data are typically only used for accident investigation purposes. NTSB told us that it was

unclear who would control and have access to streamed data and emphasized the importance of validating the data before it becomes public. As such, a few stakeholders raised privacy and security concerns with streaming flight data. The NTSB's primary concern is having access to flight data to conduct an accident investigation and NTSB officials stated that in the event of an accident in the United States, they should be the first party to download flight recorder information to ensure data integrity.[61]

- **Does not mitigate the need to recover the aircraft wreckage and fuselage, or human remains:** Identical to the concern raised above with deployable recorders, investigators still want to recover the aircraft wreckage and fuselage in order to determine the cause of the accident. Therefore, there would still be some cost for an underwater search and recovery even if investigators had the streamed FDR data.
- **Equipage and data costs:** Stakeholders, including FAA, airframe manufacturers, a trade association, and a U.S. domestic airline, stated that streaming flight recorder data could result in data transmission costs to airlines, especially using satellite communications over oceanic regions, and might require more equipment on aircraft that comes at a cost. The cost of the FLYHT's Automated Flight Information Reporting System described above is estimated at $70,000 per unit for parts and installation, according to the manufacturer. Assuming that as an average cost, we found that the total cost for the current U.S. long-haul transoceanic fleet could approach $35 million. As noted above, however, extrapolating the current per-unit cost estimate to estimate the cost for the entire fleet has certain limitations. Furthermore, it would cost between $5—$10 per minute for data streaming during emergencies, according to the manufacturer.

The FAA and other stakeholders told us that most aircraft could gather similar data using existing systems, so adding additional equipment to gather and transmit such information may not have an operational benefit for airlines. Representatives from one U.S. domestic airline that we spoke with told us that they did not see the need to equip their fleet with a new device since their system for monitoring aircraft performance could provide engine information through ACARS. Similarly, one airframe manufacturer representative told us that ACARS has helped provide information for accident investigations before FDR recovery. As noted above, ACARS, which is equipped on all new aircraft, can be programmed to transmit some aircraft operations data, including position information, when certain triggers are met.[62]

ICAO and NTSB Have Endorsed Cockpit Recorder Enhancements to Aid Accident Investigations, Though Privacy Issues Are Unresolved

ICAO announced support for extending the duration of the audio captured by the CVR, and NTSB has reiterated its recommendation for installing a cockpit image recorder. According to the NTSB, enhanced cockpit recorders would provide investigators with more information during accident investigations. Despite the potential value of this information to accident investigation, concerns over privacy remain unresolved.

Increasing the Recording Time of the Cockpit Voice Recorder

The ICAO Second High-Level Safety Conference recognized the need to increase CVR recording time to ensure that accident investigators had all relevant flight data. The concern is that CVRs should record and retain audio data for more than 2 hours given the possibility that MH370's CVR, which was designed to record on a continuous 2-hour loop, recorded over critical events in the plane's presumed 7-hour flight that could help accident investigators.[63] According to a representative from one avionics manufacturer, 2 hours is insufficient and voice recordings should cover the full duration of the flight like the FDR. Several manufacturers that we spoke with stated that they could make CVRs with additional recording time. One trade association representative told us that the CVR is meant to supplement the information recorded by the FDR and cannot definitively tell investigators what happened by itself and, therefore, cautioned that enhancing subjective audio data may not be necessary if it provides only a marginal improvement.

Installing a Cockpit Image Recorder to Aid Accident Investigation

The NTSB first issued safety recommendations for cockpit image recorders in 2000 and again in January 2015, citing a lack of valuable cockpit information during the investigations of several aviation accidents.[64] Adding a cockpit image recorder to commercial aircraft in addition to existing flight recorders could provide investigators with more information during accident investigations according to the NTSB,[65] though other stakeholders raised concerns about the subjective interpretation of images, and security and privacy issues regarding misuse of recorded images. Image recorders could record video of all the flight crew's work areas, including the instruments and controls, and provide visual documentation of the pilots' actions to the NTSB for accident investigations. FAA told us that it has not taken action because the equipage costs exceed the benefits given the extremely low aviation accident rate and the volume of data already collected by the FDR and CVR. Representatives from one trade association told us that they support the idea of providing accident investigators with more information, but they raised concerns that video, similar to audio data, is a subjective and less precise means of information gathering than FDR data. Therefore, they cautioned that video from an image recorder could lead to misinterpretation of the situation. They also cited privacy concerns with video information if improperly disclosed. The NTSB acknowledged the privacy issues with recording images of pilots in its initial recommendation, but also stated that given the history of complex accident investigations and the lack of crucial cockpit environment information, the safety of the flying public must take precedence.

CONCLUDING OBSERVATIONS

In response to the AF447 and MH370 disasters, the international aviation community is considering short- and long-term steps to improve aircraft tracking and flight data recovery with the goal of enhancing accident investigation and aviation safety. Numerous technologies—including communications systems onboard most commercial aircraft today, flight recorders that deploy from aircraft, devices capable of streaming flight recorder data in real time, and global satellite surveillance systems under development—have the potential to enhance aircraft tracking and expedite flight data recovery in the event of an oceanic accident, and industry continues to develop solutions for these tasks. However, some industry stakeholders we spoke with cited concerns about these technologies, such as the cost to equip the fleet and safety implications. Additionally, as AF447 and MH370 make clear, global air traffic control's preparedness and capability to effectively monitor oceanic flights and provide timely alerts in exceptional situations are important elements in the debate. At the international level, ICAO is finalizing its Global Aeronautical Distress and Safety System based on input from the 2015 High-Level Safety Conference, with formal adoption targeted for 2016. These developments may represent the foundation for a comprehensive, global approach to ensure that the location of an aircraft during all phases of its flight is known to authorities. However, stakeholders expressed concerns that international standards could prescribe adoption of certain solutions, such as deployable recorders. They preferred a performance-based approach that encourages voluntary adoption because of the flexibility such an approach affords industry in an era of rapidly evolving technology. Additionally, the safety record in the National Airspace System may make it difficult to demonstrate that the benefits of new equipage on U.S. airlines outweigh the costs as part of a regulatory analysis. Given the scope of the international processes that are underway, we are not making any recommendations to FAA or NTSB related to aircraft tracking or flight data recovery at this time. Ultimately, we believe it is important for FAA to remain active through the ICAO process to ensure that any new international standard for aircraft tracking and flight data recovery is consistent with a performance-based approach and is implemented in a globally harmonized manner.

AGENCY COMMENTS

We provided a draft of this report to DOT and NTSB for review and comment. Both DOT and NTSB provided technical comments that we incorporated as appropriate.

Gerald L. Dillingham, Ph.D.
Director, Physical Infrastructure Issues

APPENDIX I: ORGANIZATIONAL AFFILIATIONS OF AVIATION INDUSTRY STAKEHOLDERS GAO INTERVIEWED

Federal agencies
Federal Aviation Administration
National Transportation Safety Board
International organizations
Flight Safety Foundation
International Civil Aviation Organization
Academic institution

Embry-Riddle Aeronautical University
Airframe manufacturers
Airbus
Boeing
Airlines
Alaska Airlines
Southwest Airlines
United Airlines
Avionics manufacturers
DRS Technologies
FLYHT Aerospace Solutions Ltd.
Honeywell
L-3 Communications
Air transport communications service providers
Rockwell Collins
SITA
Industry trade associations
Air Line Pilots Association, International
Airlines for America
International Air Transport Association
Satellite communications companies
Aireon
Inmarsat

Source: GAO. | GAO-15-443

APPENDIX II: PROACTIVE USE OF FLIGHT DATA BY FEDERAL AGENCIES AND INDUSTRY

This appendix contains information describing how federal agencies and the aviation industry use operational flight data. As we reported in 2010, federal agencies and aviation industry stakeholders gather and analyze aviation data for a variety of purposes.[1] Federal agencies, including the Federal Aviation Administration (FAA) and the National Transportation Safety Board (NTSB), gather and analyze aviation data primarily to improve safety. In addition, as we reported in 2010, the aviation industry gathers quantitative and narrative data on the performance of flights and analyzes these data to increase safety, efficiency, and profitability. Aviation industry stakeholders are required to report some data to FAA— such as data on accidents, engine failures, and near midair collisions— and they have agreements with FAA and other agencies to share other data voluntarily.

For decades, the aviation industry and federal regulators, including FAA, have used data reactively to identify the causes of aviation accidents and incidents. In recent years, FAA has shifted to a more proactive approach to using data to manage aviation safety risk. The FAA continues to use data to analyze past accidents and incidents, and is also using data proactively to search for risks and address potential concerns in the National Airspace System (NAS) before accidents occur or to improve NAS efficiency. According to FAA officials, there is more safety data than ever before, and these data provide the agency with the opportunity to be more proactive about safety. FAA also recognizes that today's aviation safety, information-sharing environment is not adequate to meet the next generation needs of the NAS. According to FAA, capabilities need to be developed that can continuously extract operationally significant, safety-related information from large and diverse data sources; identify anomalous events or trends; and fuse relevant information from all available sources.

The FAA and the aviation industry have sought additional means for addressing safety problems and identifying potential safety hazards. The FAA has developed a number of programs to encourage the voluntary sharing and industry-wide analysis of operational flight data. In addition to enabling a more proactive approach to addressing safety concerns in the NAS, according to FAA, these programs could potentially enable it to predict the situations under which accidents could occur and take actions to help prevent them before occurring.

- **Flight Operational Quality Assurance (FOQA):** FOQA is a voluntary safety program designed to improve aviation safety through the proactive use of recorded flight data. Operators use these data to identify and correct deficiencies in all areas of flight operations, according to FAA. FAA officials told us that if properly used, FOQA data can reduce or eliminate safety risks, as well as minimize deviations from regulations. Through access to de-identified aggregate FOQA data, FAA can identify and analyze national trends and target resources to reduce operational risks in the NAS, air traffic control, flight operations, and airport operations, according to FAA. The value of FOQA programs, according to FAA, is the early identification of adverse trends, which, if uncorrected, could lead to accidents.

FOQA is a program for the routine collection and analysis of flight data generated during aircraft operations. FOQA programs provide more information about, and greater insight into, the total flight operations environment. FOQA data are unique because they can provide objective information that is not available through other methods, according to FAA. FAA officials told us that a FOQA program can identify operational situations in which there is increased risk, allowing the airline to take early corrective action before that risk results in an incident or accident. For example, according to representatives from one airline, FOQA analysis and findings are incorporated into flight crew training as well as airline policies and procedures. The FOQA program is another tool in the airlines' overall operational risk assessment and prevention programs. As such, according to the FAA, it must be coordinated with the airlines' other safety programs, such as the Aviation Safety Action Program and pilot reporting systems, among others.

- **Aviation Safety Information Analysis and Sharing (ASIAS):** FAA and the aviation community have initiated a safety analysis and data sharing collaboration to proactively analyze broad and extensive data to advance aviation safety, according to FAA. The initiative, known as ASIAS, leverages internal FAA data, airline proprietary safety data, publicly available data, manufacturers' data, and other data. FAA officials told us that the airline safety data are safeguarded by the MITRE Corporation, a federally funded research and development center, in a de-identified manner to foster broad participation and engagement. According to FAA, ASIAS fuses various aviation data sources in order to proactively identify safety trends and assess the impact of changes in the aviation operating environment.

ASIAS resources include both public and non-public aviation data. Public data sources include, but are not limited to, air traffic management data related to procedures, traffic, and weather. Non-public sources include de-identified data from air traffic controllers and aircraft operators, including recorded flight data and safety reports submitted by flight crews and maintenance personnel. According to FAA, governance agreements with participating airlines and owners of specific databases provide ASIAS analysts with access to safety data. Governed by a broad set of agreements, ASIAS has the ability to query millions of flight records and de-identified reports via a secure communications network, according to FAA.

Under the direction of the ASIAS Executive Board, which includes representatives from government and industry, ASIAS conducts studies, safety assessments, risk monitoring, and vulnerability discovery. In the interest of enhancing aviation safety, the results of these analyses are shared with the ASIAS participants, according to FAA. ASIAS has also established key safety benchmarks so that individual airlines may assess their own safety performance against the industry as a whole. According to aviation industry stakeholders we spoke with, the key benefit of participating in this program for the airlines is the opportunity to benchmark their individual performance against the aggregate performance of the industry. Furthermore, according to FAA, ASIAS serves as a central conduit for the exchange of data and analytical capabilities among program participants. The ASIAS vision is a network of at least 50 domestic and international airlines over the next few years, making it the only such center of its kind in the world.

- **System Safety Management Transformation (SSMT):** This FAA effort, which uses ASIAS and other data, seeks to move FAA from a post-hoc, reactive assessment of aviation safety to a more predictive, risk-assessment process. According to FAA, the SSMT project is developing data analysis and modeling capabilities that will enable FAA analysts to both determine how NextGen-related operational improvements will affect safety and evaluate potential risk mitigation strategies. One of the safety analysis methodologies the SSMT team is developing is called the Integrated Safety Assessment Model. The goal of this model is to 1) provide a risk baseline against which to measure future system changes and 2) forecast the risk and safety impacts of implementing changes to the NAS, including FAA's NextGen initiative. According to FAA, the model has been published, but continues to evolve and efforts are currently under way to refine and update its various components. The model is available as a web-based platform accessible through a user login account. According to FAA, the goal of this work is to describe in a standard format the causes and consequences of aviation accidents since 1996 as well as to describe the precursor events that contributed to these accidents. Using event sequence diagrams, FAA can describe the sequence of events led to an accident. According to FAA, current results indicate positive trends in overall impact as a means to analyze and assess baseline risks as well as emerging risks.

Airlines and airframe manufacturers also use flight data, but primarily to improve the efficiency of operations and increase profitability. As described above, domestic airlines use data collected through FOQA programs to enhance operations and proactively address maintenance issues. Airlines also use data provided through the Aircraft Communications Addressing and Reporting System (ACARS), a communications system that transmits short text messages via radio or satellite, to monitor aircraft and engine performance. According to representatives from one airline we spoke with, these data can show when certain systems and equipment need repair, and help the airline to schedule repairs in order to keep the plane in service. Another airline attributed their very low in-flight engine failure rate, in part, to proactive analysis of airplane health data. Aviation stakeholders said that this type of analysis can help identify the precursors to engine failure and help the airline address problems before the entire aircraft has to be taken out of service.

Finally, airframe manufacturers, such as Airbus and Boeing, have also developed airplane health management programs, which are offered as a service to the airlines. According to Boeing, Airplane Health Management gives airlines the ability to monitor airplane systems and parts and to interactively troubleshoot issues while the plane is in flight. Data collected through these types of programs are captured in flight and transmitted in real time to the airline's ground operations. Airbus representatives told us that their program is also designed to collect information from various aircraft systems and determine probabilities and likelihood of equipment failure. According to Boeing, airlines can use this service to make maintenance decisions before the plane has landed and be ready for any needed repairs as soon as the airplane arrives at the gate. This information is used by airlines to support operational decisions to "fix-orfly," which result in reduced schedule interruptions and increased maintenance and operational efficiency, according to Boeing. Boeing's Airplane Health Management was first introduced with the 747 and 777 aircraft models, but it has been thoroughly embedded in the design of the 787 model. To support the Airplane Health Management service for the 787, Boeing has a control center where each plane is tracked and its systems are monitored. Similarly, Airbus representatives said their airplane health management program supports the most recent aircraft models.

APPENDIX III: LIST OF COMMERCIAL AVIATION ACCIDENTS AND RECOVERY OVER WATER FROM 2000—2015

No.	Accident Date	Flight Name/Number	Aircraft Type	Accident Location	Phase	Flight Data Recorder Recovery Time	Cockpit VoiceRecorder Recovery Time
1	2/4/2015	TransAsia Airways Flight GE235	ATR-72	Keelung River, Taipei, Taiwan	Take-off	1 day	1 day
2	12/28/2014	AirAsia Indonesia QZ8501	Airbus A320	Karimata Strait, Java Sea (off the coast of Pangkalan Bun, Indonesia)	En-Route	16 days	17 days
3	3/8/2014	Malaysia Airlines Flight 370	Boeing B777	Southern Indian Ocean west of Australia (Presumed)	En-Route	Search underway	Search underway
4	1/25/2010	Ethiopian Airlines Flight 409	Boeing B737	Mediterranean Sea (Beirut, Lebanon)	Take-off	14 days	22 days
5	6/30/2009	Yemenia Flight 626	Airbus A310	Moroni, Comoros Islands (off the coast of West Africa)	Approach	60 days	60 days
6	6/1/2009	Air France Flight 447	Airbus A330	Atlantic Ocean	En-Route	1 year 11 months 2 days	1 year 11 months 3 days
7	1/15/2009	US Airways Flight 1549	Airbus A320	New York, USA (Hudson River)	Climb	7 days	7 days

No.	Accident Date	Flight Name/Number	Aircraft Type	Accident Location	Phase	Flight Data Recorder Recovery Time	Cockpit Voice Recorder Recovery Time
8	8/9/2007	Air Moorea Flight 1121	DHC6	Off the coast of Moorea, French Polynesia	Approach	N/A	21 days
9	1/1/2007	Adam Air Flight 574	Boeing B737	Makassar Strait (off the coast of Sulawesi, Indonesia)	En-Route	240 days	240 days
10	5/2/2006	Armavia Air Flight 967	Airbus A320	Black Sea (off the coast of Sochi, Russia)	Approach	22 days	20 days
11	8/6/2005	Tuninter Flight 1153	ATR-72	Off the coast of Palermo, Italy	En-Route	24 days	23 days
12	5/25/2002	China Airlines Flight 611	Boeing B747	Taiwan Strait (northeast of Makung, Penghu Islands)	Climb	25 days	24 days
13	5/7/2002	China Northern Airlines Flight 6163	MD-82	Bay near Dalian, China		14 days	7 days
14	8/23/2000	Gulf Air Flight 072	Airbus A320	Arabian Gulf near Muharraq, Bahrain	Approach	1 day	1 day
15	1/31/2000	Alaska Airlines Flight 261	MD-83	Pacific Ocean (north of Anacapa Island, California)	En-Route	3 days	2 days
16	1/30/2000	Kenya Airways Flight 431	Airbus A310	Abidjan (Off the Ivory Coast, West Africa)	Take-off	6 days	26 days

Source: GAO analysis of Boeing and Bureau d'Enquêtes et d'Analyses data. | GAO-15-443.

End Notes

[1] For example, see Bureau d'Enquêtes et d'Analyses, Final Report on the Accident on 1st June 2009 to the Airbus A330-203 registered F-GZCP operated by Air France Flight AF 447 Rio de Janeiro – Paris (France: July 2012) and Malaysian Ministry of Transport, The Malaysian ICAO Annex 13 Safety Investigation Team for MH370, Factual Information Safety Investigation for MH370 (Malaysia: March 2015).

[2] National Transportation Safety Board (NTSB), Safety Recommendation A-15-1 through - 8, (Washington, D.C.: Jan. 22, 2015). See also NTSB, Safety Recommendation A-00-30 and - 31 (Washington, D.C.: Apr. 11, 2000) and Transportation Safety Board of Canada, Reassessment of the Responses to Aviation Safety Recommendation A03-08 (Canada: March 2003).

[3] For example, see ICAO Ad-Hoc Working Group on Aircraft Tracking, Concept of Operations: Global Aeronautical Distress and Safety System (GADSS), Final Draft Version 4.1 (October 2014) and International Air Transport Association Aircraft Tracking Task Force, Report and Recommendations (November 2014).

[4] For example, see DRS Technologies, Honeywell, FLYHT Aerospace Solutions, Ltd, and L-3 presentations to NTSB Forum, Emerging Flight Data and Locator Technology (Washington, D.C.: Oct. 7, 2014). See also Inmarsat, Iridium, and Aireon presentations to ICAO Multidisciplinary Meeting Regarding Global Tracking (May 13, 2014).

[5] Honeywell, Future Air Navigation System, accessed June 20, 2014, https://aerospace.honeywell.com/news/mandates-fans-1a and Universal Avionics Systems Corporation, Understanding the Future Air Navigation System (FANS) 1/A Operations and Regulatory Requirements, White Paper Doc No.: WHTP-2013-18-10 (October 2013).

[6] The National Airspace System (NAS) includes air traffic control systems, air traffic control procedures, operational facilities, aircraft, and the people who certify, operate, and maintain them. The NAS covers U.S. airspace, which generally extends 12 nautical miles from the coastline.

[7] The Convention on International Civil Aviation, commonly known as the "Chicago Convention," was signed on December 7, 1944 by 52 nations. In October 1947, ICAO became a specialized agency of the United Nations. There are currently 191 parties to the Convention.

[8] Radar is a method whereby radio waves are transmitted into the air and are then received when they have been reflected by an object in the path of the beam. Range is determined by measuring the time it takes (at the speed of light) for the radio wave to go out to the object and then return to the receiving antenna. The direction of a detected object from a radar site is determined by the position of the rotating antenna when the reflected portion of the radio wave is received. U.S. Department of Transportation, Federal Aviation Administration, Aeronautical Information Manual: Official Guide to Basic Flight Information and ATC Procedures (Washington, D.C.: Apr. 3, 2014), accessed March 30, 2015, http:/www.faa.gov/atpubs.

[9] Some portions of the NAS have upgraded to ADS-B for surveillance. As specified in 14 C.F.R. § 91.225, the FAA has mandated that aircraft operating in specified airspace within the NAS be equipped for ADS-B by January 1, 2020.

[10] ICAO requires that, at a minimum, aircraft operating over oceans must have a functioning two-way radio to communicate with the appropriate air traffic control unit. FAA requires Part 121 operators (i.e., scheduled commercial air carriers) to carry certain communication and navigation equipment for extended over-water operations. For example, aircraft of these

operators must have at least two independent long-range navigation systems and at least two independent long-range communication systems to communicate with at least one appropriate station from any point on the route.

[11] Designated reporting points are certain points during flight where pilots are required to send position information, as indicated by symbols on en route aeronautical charts. U.S. Department of Transportation, Federal Aviation Administration, Aeronautical Information Manual: Official Guide to Basic Flight Information and ATC Procedures (Washington, D.C.: Apr. 3, 2014), accessed March 30, 2015, http:/www.faa.gov/atpubs.

[12] FAA, Advisory Circular 91-70A, Oceanic and International Operations (Washington, D.C.: Aug. 12, 2010).

[13] FANS was developed by ICAO in partnership with Boeing, Airbus and others in the air transport industry. Today, FANS-1 is used on Boeing aircraft, while the Airbus version is known as FANS-A. In this report, we refer to the terms generically as FANS.

[14] Part 121 prescribes rules governing the domestic, flag, and supplemental operations to hold an air carrier certificate. Scheduled-service airlines are generally issued a Part 121 certificate by FAA and operate turbojet-powered airplanes, airplanes with more than nine passenger seats, or airplanes having a payload capacity of more than 7,500 pounds.

[15] For example, certificate holders conducting operations with airplanes with more than 30 passenger seats or having a payload capacity of more than 7,500 pounds are required to have a flight following system.

[16] Annex 12 of ICAO's Chicago Convention recommends that search and rescue regions coincide with flight information regions when practicable.

[17] This program is an intergovernmental cooperative of 41 countries and two agencies. Its mission is to provide accurate, timely, and reliable distress alert and location data to help search and rescue authorities assist persons in distress. The program provides alerts to over 200 countries and territories even if they are not a member.

[18] While ICAO standards call for commercial passenger aircraft to be equipped with an emergency locator transmitter, emergency locator transmitters are not required of scheduled flights of commercial passenger aircraft operating in the U.S. See 49 U.S.C. § 44712(c)(1)(A) and 14 C.F.R. § 91.207(f)(2).

[19] Article 26 of ICAO's Chicago Convention provides that the country in which an accident or incident occurs is responsible for instituting the investigation into its circumstances in so far as its laws permit, but Annex 13 provides that such country may delegate the entirety or part of the investigation to another country by mutual arrangement or consent. Under Annex 13, when the location of the accident or incident cannot be definitely established as being in the territory of any country, responsibility for the investigation falls to the country of the aircraft's registry.

[20] The NTSB participates in the investigation of aviation accidents and serious incidents outside the United States pursuant to Annex 13 to the Convention on International Civil Aviation.

[21] Data recorded by aircraft systems may also be used in airline Flight Operational Quality Assurance (FOQA) programs. According to FAA, FOQA programs are designed to make commercial aviation safer by allowing commercial airlines and pilots to share de-identified aggregate information with FAA so that FAA can monitor national trends in aircraft operations and target its resources to address operational risk issues (e.g., flight operations, air traffic control, and airports). Data from FOQA and other sources are used in the Aviation Safety Information and Analysis Sharing (ASIAS) initiative, a collaborative effort between FAA and the aviation community to proactively analyze data to enhance aviation safety. See appendix II for additional information.

[22] The standards in Annex 6 call for an increase in the cockpit voice recording time from 30 minutes to 2 hours by January 2016. The FAA already requires at least a 2-hour cockpit voice recording duration.

[23] FAA, Advisory Circular 91-70A.

[24] Coordinated Universal Time is the 24-hour time standard common to every place in the world. Formerly and still widely called Greenwich Mean Time, UTC nominally reflects the mean solar time along the Earth's prime meridian.

[25] Australian Transport Safety Bureau, Considerations on defining the search area (May 26, 2014), accessed February 19, 2015, http://www.atsb.gov.au/publications/2014/considerations-on-defining-the-search-areamh370.aspx.

[26] Typically, the aircraft transmits a "log-on" request in order to connect to the satellite communications system and which is acknowledged by the ground station. Once connected, if the ground station has not heard from the aircraft within an hour, it will check that the connection is still operational by transmitting a "log on interrogation" using the aircraft's unique identifier. If the aircraft receives this information, it returns a short message—the handshake—that it is still logged on to the network.

[27] The satellite communication system consisted of the Inmarsat Classic Aero ground station in Perth, Western Australia and the Inmarsat Indian Ocean Region I-3 satellite.

[28] Australian Transport Safety Bureau, MH370 – Definition of Underwater Search Areas, AE-2014-054 (June 26, 2014), updated Aug. 18, 2014.

[29] Ministry of Transport, Malaysia. Malaysian ICAO Annex 13 Safety Investigation Team for MH370, Factual Information: Safety Investigation for MH370 (Mar. 8, 2015).

[30] IATA Aircraft Tracking Task Force, Report and Recommendations (Nov. 11, 2014).

[31] According to ICAO, the responsibility for aircraft tracking lies with the commercial airline operator.

[32] In 2010, the Single European Sky Air Traffic Management Research Joint Undertaking, which is a public-private partnership responsible for research, development and validation of technology and procedures relating to European air traffic modernization, launched the Oceanic Position Tracking Improvement and Monitoring Initiative as a collaborative project with air navigation service providers, airlines, manufacturers, satellite communication providers, and other entities involved in the aviation sector in European oceanic airspace.

[33] The Task Force proposed that any aircraft tracking system should have the capability to report at a faster rate based on established triggering parameters, but did not specify what that faster reporting rate should be.

[34] The flight management system provides the primary navigation, flight planning, and optimized route determination for the aircraft.

[35] According to company representatives, the Automated Flight Information Reporting System has been installed on over 30 scheduled and charter airlines and on 10 business jets, special mission, and military operators worldwide. First Air of Canada has fully implemented the system, which also has the capability to stream up to all flight data recorder parameters during emergencies.

[36] According to one major air transport communications service provider, the major freight delivery airlines are also equipped with ACARS. However, ACARS installation and usage may not be justified for smaller commercial passenger aircraft operators.

[37] For example, one component of FAA's NextGen initiative is Data Communications (Data Comm), which envisions increasing controller to flight crew communications enabled by advanced communication technologies, including FANS avionics.

[38] The North Atlantic airspace utilizes a constantly changing 12-hour "track" or flight path system designed around the high altitude winds and weather to optimize flights each day. Because there are over 1,400 aircraft crossing the North Atlantic each day and more traffic expected in the coming years, air traffic control has mandated a phased approach that gradually requires FANS equipage in order to increase airspace capacity and enhance safety. Mandates for FANS began in 2013, and by 2020, the entire North Atlantic airspace will require aircraft to be FANS-equipped.

[39] There could also be a cost up to $100,000 to activate FANS in a flight management system with the capability not activated, according to one air transport communications service provider. In addition to the FANS equipment needed on board the aircraft, air traffic control also needs to be equipped with FANS systems on the ground to receive the FANS communications from the aircraft. According to one air transport communications service provider we spoke to, about 50 of the total 160 air navigation service providers worldwide have FANS ground systems.

[40] The Automated Flight Information Reporting System already includes satellite communications capability, according to company representatives.

[41] Typically, ADS-C messages are exchanged between aircraft and air traffic control. The SITA Aircom Server Flight Tracker would provide those ADS-C messages to airline operators so that the airlines could adapt the pace of flight tracking as needed. According to company representatives, this flight tracking proposal is still under development.

[42] According to Rockwell Collins, the ARINC Multilink flight tracking service brings together multiple data sources to report the location of an aircraft anywhere in the world. The use of multiple sources of data, according to the company, means an aircraft's position can be reported more frequently. Other flight tracking products are available from a variety of commercial vendors.

[43] Through its NextGen initiative, FAA is implementing a major redesign of the air transportation system in the United States to increase efficiency, enhance safety, and reduce flight delays. NextGen is planned to incorporate precision satellite navigation and surveillance; digital, networked communications; an integrated weather system, and more. The implementation of NextGen requires the involvement of airlines and other aviation stakeholders, since those entities will need to invest in new avionics and other technologies to take advantage of NextGen technologies.

[44] Aireon proposes a space-based global aircraft surveillance system to provide both airlines and air traffic service providers with very specific aircraft location position information. This system could provide global surveillance of aircraft, including remote and oceanic areas, using satellite technology.

[45] On December 16, 2014, the Consolidated and Further Continuing Appropriations Act, 2015 (Pub. L. No. 113-325, 128 Stat. 2130 (2014)) was enacted into law. The Explanatory Statement accompanying this appropriations act directs $7.5 million to advance the use of space-based ADS-B for air traffic control separation services, support the collection and validation of surveillance data, and help assess the impact on FAA's oceanic automation system. The Explanatory Statement directs the FAA to make an investment decision regarding satellite-based ADS-B no later than 30 days after the enactment of the appropriations act. FAA is studying the feasibility of space-based ADS-B.

[46] According to company representatives, the Aireon ALERT service will be managed from the Irish Aviation Authority's North Atlantic Communications Center located on the West Coast of Ireland. Aireon ALERT will make it possible to query the location and last flight path of any ADS-B equipped aircraft flying in airspace beyond the reach existing radar and

other terrestrial surveillance systems. According to Aireon, the data would be provided free of charge in near real-time to air navigation service providers, airlines and search and rescue authorities, even if they are not customers of Aireon's core space-based ADS-B service.

[47] ICAO Ad-Hoc Working Group on Aircraft Tracking, Concept of Operations: Global Aeronautical Distress & Safety System (GADSS), Final Draft Version 4.1 (October 2014).

[48] In January 2015, NTSB recommended that all aircraft used in extended overwater operations and under Title 14 CFR Part 121 or Part 135 that are required to have a CVR and FDR be equipped with a tamper-resistant method to broadcast to a ground station sufficient information to establish the location where an aircraft terminates flight as the result of an accident within 6 nautical miles of the point of impact. NTSB, Safety Recommendation A-15-1 through -8 (Washington, D.C.: Jan. 22, 2015).

[49] According to ICAO, the performance specifications for the in-flight triggering criteria and broadcasting rate to be used are still under development.

[50] Executive Order 12866, Regulatory Planning and Review, 58 Fed. Reg. 51735 (Washington, D.C.: Oct. 4, 1993) directs each federal agency proposing a regulation to make a reasoned determination that the benefits of the intended regulation justify its costs. See also Office of Management and Budget, Regulatory Analysis, OMB Circular A4 (Washington, D.C.: Sept. 17, 2003).

[51] System Wide Information Management consists of standards, infrastructure, and governance enabling air traffic management information and its exchange between qualified parties via interoperable services, according to ICAO.

[52] A TSO is a minimum performance standard for specified materials, parts, and appliances used on civil aircraft. When authorized to manufacture a material, part, or appliance to a TSO standard, it is called a TSO authorization. Receiving a TSO authorization is both design and production approval. Receiving a TSO authorization is not an approval to install and use the article in the aircraft. It means that the article meets the specific TSO and that the applicant is authorized to manufacture it.

[53] 80 Fed. Reg. 12698 (Mar. 10, 2015).

[54] In a March 2012 amendment to Annex 6 Part I, ICAO stated that not later than January 1, 2018, ULBs shall operate for a minimum of 90 days.

[55] See NTSB, Safety Recommendation A-15-1 through -8 (Washington, D.C.: Jan. 22, 2015).

[56] FAA granted TSO design approval for an automatic deployable flight recorder on June 10, 1999.

[57] Accident locations include seas, gulfs, rivers, bays, straits, and oceanic regions. See appendix III for a more detailed description of location for each accident identified.

[58] We reviewed data included in Boeing's presentation at the NTSB Forum, Emerging Flight Data and Locator Technology (Washington, D.C., Oct. 7, 2014) and the BEA, Flight Data Recovery Working Group Report, accessed July 28, 2014, http://www.bea.aero/en/enquetes/flight.af.447/flight.data.recovery.working.group.final.repo rt.pdf. We then limited our analysis to commercial aviation accidents that occurred over water from 2000 forward, up to and including the Air Asia QZ8501 accident from December 28, 2014 and the TransAsia Airways Flight GE235 accident, which crashed into the Keelung River in Taiwan shortly after takeoff on February 4, 2015.

[59] According to one manufacturer of deployable recorders, line items include cost of the actual deployable recorder system (less than $30,000 each), aircraft wiring and mechanical installation kits, installation labor, and non-recurring engineering costs. Specifically, the non-recurring costs relate to engineering and installation design, qualification, and certification for obtaining FAA Supplemental Type Certification (STC) for the first aircraft

"type" for both foreign and U.S. made aircraft that a U.S. airline may purchase. There may be additional costs for each aircraft type model (B737-200, -300) if there are differences in the avionics configuration to the deployable recorder.

[60] Legislation has been introduced in Congress designed to improve flight recorder and aircraft location requirements on certain passenger aircraft. For example, H.R. 5337, the Safe Aviation and Flight Enhancement Act of 2014, which called for a deployable combination flight data and cockpit voice recorder with an emergency locator transmitter as a second combination flight recorder system installed on commercial passenger aircraft ordered by an air carrier on or after January 1, 2016, was introduced on July 31, 2014, but was not passed by the House of Representatives. H.R.5337, 113th Cong. (2nd Sess. 2014). In the current session of Congress, a similar bill, H.R. 772, the Safe Aviation and Flight Enhancement Act of 2015, was introduced in the House of Representatives on February 5, 2015. H.R. 772, 114th Sess. Cong. (1st Sess. 2015).

[61] Outside of the United States, other government accident investigators may be the first party to download the flight recorder data.

[62] For example, AF447 transmitted a position message and a series of 24 maintenance messages via ACARS during its final minutes of flight. Investigators were able to use this information to determine the flight's last known position and as a starting point for the accident investigation.

[63] These events include air traffic control's loss of contact with the flight crew and the aircraft's deviation from its approved flight plan.

[64] NTSB, Safety Recommendation, A-00-30 and -31 (Washington, D.C.: Apr. 11, 2000); NTSB, Safety Recommendations, A-15-1 through -8 (Washington, D.C.: Jan. 22, 2015).

[65] In a recent helicopter accident report, NTSB noted that images captured by the helicopter's onboard image recorder significantly aided the investigation by providing valuable insight into the pilot's actions, and as a result, was able to develop safety recommendations to help prevent future accidents. See NTSB, Aircraft Accident Report: Crash Following Encounter with Instrument Meteorological Conditions After Departure from Remote Landing Site, Alaska Department of Public Safety, Eurocopter AS350 B3, N911AA, Talkeetna, Alaska, March 30, 2013. NTSB/AAR-14/03/PB2014-108877 (Washington, D.C.: Nov. 5, 2014).

End Note for Appendix II

[1] GAO, Aviation Safety: Improved Data Quality and Analysis Capabilities are Needed as FAA Plans a Risk-Based Approach to Safety Oversight, GAO-10-414 (Washington, D.C.: May 6, 2010).

Chapter 2

SAFETY RECOMMENDATION A-15-1 THROUGH -8[*]

Christopher A. Hart

The Honorable Michael P. Huerta
Administrator
Federal Aviation Administration
Washington, DC 20590

The National Transportation Safety Board (NTSB) has long been concerned about rapid recovery of recorded information to guide investigations, help determine accident causes, and develop recommendations to prevent recurrences. To focus attention on this issue, the NTSB convened its *Emerging Flight Data and Locator Technology Forum* on October 7, 2014, in Washington, D.C.[1] Forum discussions among government, industry, and investigative experts helped identify the following safety issues:

- The need for improved technologies to locate aircraft wreckage and flight recorders following an accident in a remote location or over water
- The need for timely recovery of critical flight data following an accident in a remote location or over water

---

[*] This is an edited, reformatted and augmented version of a letter issued by the National Transportation Safety Board, January 22, 2015.

- This letter presents recommendations to address these safety issues and discusses open recommendations regarding cockpit image recorders and the need to protect flight recorder systems against intentional or inadvertent deactivation. The NTSB urges the Federal Aviation Administration (FAA) to take action on the safety recommendations issued in this letter.

BACKGROUND

Cockpit voice recorder (CVR) and flight data recorder (FDR) data are some of the most important information sources available to help determine causes of aviation accidents. As such, recovering the recorders is an investigative priority at the crash site. Recent events have highlighted that recovering flight data can be costly and difficult when an accident occurs in a remote area, outside of radar coverage. For example, on June 1, 2009, Air France Flight 447, an Airbus A330, crashed into the Atlantic Ocean during a regularly scheduled flight from Rio de Janeiro, Brazil, to Paris, France.[2] All 228 passengers and crew on board died. Although some wreckage was recovered during the first few days of search activity, it took almost 2 years and cost about $40 million to locate and recover the flight recorders.[3] When the recorders were finally recovered, the information they contained was essential to the French Bureau d'Enquêtes et d'Analyses (BEA) in issuing its findings and conclusions regarding the cause of the accident. Additionally, on March 8, 2014, Malaysia Airlines Flight 370 disappeared while on a scheduled flight from Kuala Lumpur, Malaysia, to Beijing, China, with 239 passengers and crew on board. According to the Malaysian Ministry of Transport's preliminary report, 26 countries have participated in the search effort using 82 aircraft and 84 vessels.[4] To date, those participating in the search have analyzed and mapped more than 41,000 square kilometers of ocean floor without locating the aircraft's CVR, FDR, or any other wreckage.[5] The investigation remains open.

AIRCRAFT POSITION REPORTING

When traveling within areas of radar coverage, aircraft transmit data to ground radar stations via transponders that allow position to be determined.

When traveling outside of radar coverage, pilots periodically communicate the position of their aircraft with Air Traffic Organizations (ATO) and airline dispatchers via a variety of methods. For example, pilots can communicate their position via radio with ATOs when passing certain waypoints on their flight plan. Many large transport category aircraft are also equipped with a digital data link system. One common digital data link system is an Aircraft Communications Addressing and Reporting System (ACARS), which can be configured to automatically report aircraft position periodically to a ground station via satellite communications.[6] Additionally, many aircraft are equipped with Automatic Dependent Surveillance-Broadcast (ADS-B) systems that periodically transmit their identification, current position, altitude, and speed to ATOs and other aircraft.[7]

During the 2 years before the Air France Flight 447 recorders were located and recovered, the BEA convened a working group to study flight data transmission, flight recorder technology, and wreckage localization technology.[8] The working group consisted of more than 120 international members from investigative bodies, aircraft manufacturers, regulatory authorities, recorder manufacturers, and satellite service providers, including the NTSB, the FAA, and US industry representatives. The working group examined information relating an aircraft's rate of transmission of position to an accident impact location for 44 accidents of transport category aircraft. In 95 percent of the cases, the working group determined that if these aircraft had been transmitting their position once every minute, the last reported position would have been within a 6 nautical mile (nm) radius of the point of impact.[9] The International Civil Aviation Organization (ICAO) Flight Recorder (FLIREC) Panel was tasked with proposing recommendations that would aid in the recovery of data from flight recorders following an accident. The FLIREC Panel is using the BEA working group results to propose amendments that will better determine the position of an accident impact location. This work is ongoing.

In the case of Air France Flight 447, the ACARS system was configured to transmit the aircraft's position about once every 10 minutes. Given the aircraft's cruising speed and altitude, this resulted in a search area with a radius of 40 nm from its last reported position. Such a large area made the search much more challenging. If the aircraft had reported its position more frequently, the search area could have been significantly reduced.

As a consequence of the desire to more quickly and efficiently locate a downed aircraft, changes have been and are being made. In 2011, Air France modified the data link communications systems on its long-haul aircraft to report position once every minute under certain conditions.[10] Further, at the October 2014 NTSB forum, the European Aviation Safety Agency (EASA) representative to the United States stated that the European Commission, assisted by EASA, is drafting additional performance-based regulatory material to improve aircraft tracking. Additionally, in May 2014, ICAO held a Multidisciplinary Meeting Regarding Global Flight Tracking to discuss the issues of aircraft tracking and accident location. ICAO established a framework for future efforts in this area, including an industry task force and the development of a flight-tracking concept of operations.[11] As a result, an ICAO working group has developed proposed requirements for a Global Aeronautical Distress and Safety System, which addresses tracking of aircraft during routine and abnormal flight conditions.[12] The NTSB supports the international efforts underway in this area and believes similar action should be taken in the United States. Therefore, the NTSB concludes that aircraft should broadcast sufficient information to facilitate a quicker identification of an accident location, a faster search and rescue response, and a more effective underwater search effort. Further, the BEA's working group analysis indicates that broadcasting an aircraft's location within 6 nm of the point of impact is realistically achievable with modern data link technology.

Discussions at the NTSB forum indicated that there are many technologies available that would enable broadcasting the location of an accident within 6 nm. Use of an emergency locator transmitter (ELT), whether as a stand-alone unit or as part of an automatically deploying flight recorder, could transmit the location of an accident; however, it must remain above water to function. In addition, frequent broadcast of an aircraft's position might establish the location of an accident within 6 nm. Currently, many aircraft are equipped with ELTs and many are also capable of automatically transmitting their position, speed, and other information. Although the primary goal of aircraft position reporting in the event of an accident is to reduce the search area to a radius of less than 6 nm, the NTSB recognizes there are complexities associated with issuing such a requirement and implementing the technology. Some operators have already modified their data link communications systems to report aircraft position every minute under certain conditions. Further, many other aircraft are already equipped with similar technology capable of automatically reporting aircraft position. The NTSB encourages the voluntary implementation of increased frequency of position reporting by operators as an interim measure to better establish accident location following an event.

The NTSB is most concerned about aircraft that fly extended overwater (EOW) operations outside of radar coverage because timely response and recovery are more challenging when an accident occurs in a remote area.[13] Currently, aircraft that fly EOW operations must carry additional survival equipment, such as life rafts and survival-type ELTs, to mitigate the risks.[14] The NTSB recommends that the FAA require that all aircraft used in EOW operations and operating under Title 14 *Code of Federal Regulations* (CFR) (1) Part 121 or (2) Part 135 that are required to have a CVR and an FDR, be equipped with a tamper-resistant method to broadcast to a ground station sufficient information to establish the location where an aircraft terminates flight as the result of an accident within 6 nm of the point of impact.

MORE EFFECTIVE LOCATION OF UNDERWATER WRECKAGE

The search for flight recorders submerged under water has traditionally been guided by an underwater locator beacon (ULB) affixed to each recorder. ULBs transmit an ultrasonic signal, or "ping," when submerged in water. Once activated by submersion in water, the device's batteries can power this signal continuously for at least 30 days. In some newer models, the batteries can power the signal for at least 90 days. The signal's detection range is generally limited to 1 to 3 nm depending on depth, underwater topography, and surrounding conditions. The detection range also can be greatly reduced if a ULB is covered or blocked by aircraft wreckage. Delays in locating a crash site and mobilizing underwater search assets have limited the available search time while the ULB batteries are still operational.

Following the Air France Flight 447 accident, SAE International formed a working group and developed Aerospace Standard AS6254, *Minimum Performance Standard for Low Frequency Underwater Locating Devices (Acoustic) (Self-Powered)*.[15] Underwater locating devices (ULD) are beacons designed to be mounted to the fuselage structure and generate a lower frequency signal than ULBs, which are smaller and are primarily used to aid in locating the recorders within the wreckage field. The lower operating frequency of ULDs increases signal detection range and improves signal transmissibility through any aircraft structure to allow main wreckage field identification. Another advantage of ULDs is that many private and military resources have receivers capable of detecting these low frequency signals, which increases the likelihood that resources capable of searching for aircraft wreckage would be in closer proximity to the search area. In 2012, the FAA issued a new Technical Standard Order (TSO), TSO-C200, "Airframe Low Frequency Underwater Locating Devices (Acoustic) (Self-Powered)," outlining the minimum performance standard for airframe low frequency ULDs based on the SAE International AS6254 standard. In 2012, ICAO adopted the SAE International standard for ULDs on aircraft that operate long distances over water. Although the NTSB is encouraged by the development of TSO-C200 and AS6254, both define the minimum operating life of a ULD to be 30 days. The NTSB believes a ULD minimum operating life should be at least 90 days. The NTSB concludes that airframe-mounted ULDs would improve the underwater search for aircraft wreckage because of their longer detection range and lower frequency signals. The ULD low frequency aids in locating the main wreckage field and supplements the use of a ULB, which has a shorter range but a higher frequency that provides a more precise location to identify recorder position. Therefore, the NTSB recommends that the FAA require that all aircraft used in EOW operations and operating under 14 CFR (1) Part 121 or (2) Part 135 that are required to have a CVR and an FDR, be equipped with an airframe low frequency ULD that will function for at least 90 days and that can be detected by equipment available on military, search and rescue, and salvage assets commonly used to search for and recover wreckage.

SUPPLEMENTAL METHODS TO RECOVER FLIGHT DATA

The NTSB is also interested in ways to recover critical flight data in a more timely manner and that do not require immediate underwater retrieval of flight recorders. Locating and recovering flight recorders in over water accidents has been more problematic than those occurring on land.[16]

Once recovered, flight recorders have been highly reliable, and data have been successfully downloaded. However, there have been rare instances in which recorders have not been recovered or data were lost due to damage from exposure to severe fire or underwater conditions.[17] Because of this, the NTSB issued Safety Recommendation A-99-17 to the FAA requiring the installation of dual combination flight recorders that include both cockpit voice and flight data recording functionalities on board newly built aircraft.[18] Although the NTSB continues to believe that dual combination recorders provide a very effective level of information redundancy, technology advances over the past several years have yielded alternate means to provide some degree of recorded flight data redundancy without the delays associated with a difficult underwater recovery.

Deployable recorders are a technology that can be used to recover flight data without the delay of a long and expensive underwater recovery. At the October 2014 NTSB forum, the Director of Air Programs at DRS Technologies Canada Ltd. stated that deployable recorders have been used in military and over water helicopter applications since the 1960s and are currently available from several manufacturers. These recorders combine traditional FDR and CVR functions into one unit and are capable of providing a comparable amount of flight data. They are designed to separate from the aircraft upon fuselage structural deformation or when submersed in water. If in water, they float indefinitely on the surface. These units are also equipped with ELTs that operate on the 121.5 megahertz (MHz) and 406 MHz frequencies for location and recovery. Standards already exist for automatically deploying flight recorders.[19] The European Organization for Civil Aviation Equipment (EUROCAE) Document ED-112A, *Minimum Operational Performance Specification for Crash Protected Airborne Recorder Systems*, includes unique survivability criteria specific to the operational characteristics of deployable recorders with the goal of providing a comparable level of survivability to fixed flight recorders. Deployable flight recorders are subject to different impact and fire survivability requirements because they are designed to separate from an aircraft and come to rest outside of the primary wreckage field. The BEA Flight Data Recovery Working Group identified the installation of deployable recorders as a viable solution to supplement traditional flight recorders. At the NTSB forum, the Chief Product Security Officer at Airbus indicated the company's intention to study the application of deployable recorders for use in its aircraft. Installing deployable recorders, if combined with traditional onboard flight recorders, would provide an increased level of information redundancy.

Triggered flight data transmission is another promising technology, which was also studied by the BEA working group. This technology involves monitoring preselected aircraft parameters and triggering satellite transmission of critical flight data when the parameters deviate from their normal operating envelope. At the NTSB forum, the Director of Strategic Programs at FLYHT Aerospace Solutions Ltd., a manufacturer of flight data transmission technology, testified that triggered flight data transmission was not only feasible, but also already in service on some aircraft. Additionally, at this time, manufacturers and operators are equipping their aircraft with commercial satellite communications systems that can support broadband video, voice, and data transmissions. Commercial satellite systems on the market today are primarily used for passenger and crew connectivity and can support speeds of 200-400 kilobits per second (kbps). Higher speed capability is forthcoming. Such bandwidth would enable real-time parametric flight data transmission to begin after a triggering event as well as transmission of a limited amount of stored flight data recorded before the triggering event.[20] For example, once triggered, each second of flight data between the triggering event and the end of the flight could be transmitted. Additionally, any available bandwidth not required for transmitting the real-time data could be used to transmit as much data as possible from prior to the triggering event.[21]

Either deployable recorder or triggered flight data transmission technologies, if used to supplement the mandatory onboard recorders, would provide investigators more timely access to information and offer valuable insight into the circumstances near the end of an accident flight. Access to the parametric flight data is available via both technologies.[22] These data would provide information to help focus an investigation while the search for and recovery of recorders and wreckage is taking place at underwater crash sites. The NTSB concludes that supplemental methods of flight data recovery are feasible and would facilitate rapid identification of critical safety issues. Therefore, the NTSB recommends that the FAA require that all newly manufactured aircraft used in EOW operations and operating under 14 CFR (1) Part 121 or (2) Part 135 that are required to have a CVR and an FDR, be equipped with a means to recover, at a minimum, mandatory flight data parameters; the means of recovery should not require underwater retrieval. Data should be captured from a triggering event until the end of the flight and for as long a time period before the triggering event as possible.

The NTSB recognizes there are significant ongoing international industry and regulatory efforts to develop and adopt standards for enhanced aircraft position reporting and supplemental methods for recovering flight data. Achieving these goals on a global basis will demand a harmonized approach that addresses the needs of many stakeholders and ensures that domestic and foreign parties operate under equivalent standards. The NTSB concludes that the FAA should work closely with its international counterparts to ensure that regulations affecting both domestic and foreign parties are consistent. The NTSB believes that the current ICAO Standards and Recommended Practices address the intent of Safety Recommendation A-15-2 at the international level.[23] Therefore, the NTSB recommends that the FAA, concurrent with the implementation of Safety Recommendations A-15-1 and A-15-3, coordinate with other international regulatory authorities and ICAO to harmonize the implementation of the requirements outlined in Safety Recommendations A-15-1 and A-15-3.

PREVIOUS RECOMMENDATIONS ON COCKPIT IMAGE RECORDING SYSTEMS AND PROTECTION AGAINST INTENTIONAL OR INADVERTENT DEACTIVATION OF RECORDING SYSTEMS

The NTSB also reviewed previously issued safety recommendations on cockpit image recording systems and protection against deactivation of recording systems. In April 2000, citing several accidents involving a lack of information regarding crewmember actions and the flight deck environment, including ValuJet Flight 592, SilkAir Flight 185, Swissair Flight 111, and EgyptAir Flight 990, the NTSB issued Safety Recommendations A-00-30 and -31 to the FAA.[24] In addition to recommending that the FAA require placing recorder system circuit breakers in locations the flight crew could not access, Safety Recommendation A-00-30 recommended that the FAA mandate that in-service aircraft operated under 14 CFR Part 121, 125, or 135 be equipped with a crash-protected cockpit image recording system. Safety Recommendation A-00-31 recommended similar action for newly manufactured aircraft that would be operated under 14 CFR Part 121, 125, or 135. In 2006, the NTSB reiterated Safety Recommendation A-00-30 as a result of its investigation of the October 19, 2004, accident involving Corporate Airlines Flight 5966, a BAE-J3201 aircraft, in Kirksville, Missouri.[25] The NTSB believes it is now appropriate to clarify the recommendations by separating the issue of recorder system circuit breaker accessibility from the issue of cockpit image recording systems and to update the recommendations by incorporating government and industry developments in cockpit image recording systems.

In response to Safety Recommendations A-00-30 and A-00-31, concerning circuit breakers, the FAA replied that the circuit protection for any electrical system that is active during flight should be accessible to the flight crew so that in the event of an in-flight electrical fire, the crew can quickly cut power to all electrical equipment in accordance with approved procedures. However, the NTSB notes that there are often circuit breakers on advanced aircraft to which flight crews do not have access. Further, some manufacturers currently install the flight recorder circuit breakers in a location that the flight crew cannot readily access. Therefore, the NTSB believes that it is possible to incorporate protections against disabling specific equipment while also meeting applicable regulations as well as the FAA's stated design philosophy. The NTSB concludes that flight crews may disable recording systems on current transport category aircraft, which could hinder the identification of safety issues during the investigation of serious events. Therefore, the NTSB recommends that the FAA require that all newly manufactured transport category aircraft incorporate adequate protections against disabling flight recorder systems.

The NTSB recognizes that retrofitting existing aircraft to incorporate protections against disabling recording systems may present significant challenges. Rewiring an aircraft would be difficult, but there may be other possible options to ensure that recording systems cannot be easily disabled. Some possible examples to help reduce accessibility include moving the circuit breakers to a different location or adding a protective cover. Despite the challenges with protecting recording system circuit breakers on existing aircraft, the NTSB concludes that identifying potential solutions to incorporate these protections is important because of the large number of aircraft that will remain in service for many years. Therefore, the NTSB recommends that the FAA identify ways to incorporate adequate protections against disabling flight recorder systems on all existing transport category aircraft.

In its final report on the Air France Flight 447 accident, the BEA cited difficulties in reconstructing critical instrument panel indications that were available to the flight crew. Consequently, the BEA recommended that ICAO require public transport flights with passengers be equipped with a cockpit image recorder that can record the instrument panel and also that guidelines be established to guarantee the confidentiality of recordings.

On September 3, 2010, a Boeing 747-44AF, operated by United Parcel Service (UPS), crashed while attempting to return to Dubai International Airport following an in-flight cargo fire. Some critical information, such as flight instrument indications, switch positions, and aircraft system conditions, could not be confirmed with the available evidence. The final report, prepared by the United Arab Emirates General Civil Aviation Authority in accordance with ICAO Annex 13, specifically cited the lack of cockpit imagery as a detriment to the timeliness of the investigation and delivery of critical safety recommendations.[26]

In the SilkAir and EgyptAir crashes, the CVR and the FDR recordings provided limited information about crew actions and the status of the cockpit environment. Further, in the Air France and UPS crashes, the accident aircraft were equipped with FDRs that greatly exceeded the minimum parameter requirements. However, in these accidents, critical information related to the cockpit environment conditions (for example, crew actions and visibility), instrument indications available to crewmembers, and the degradation of aircraft systems was not available to investigators. The NTSB concludes that image recordings would provide critical information about flight crew actions and the cockpit environment that has not been provided by CVRs and FDRs, and would enhance accident investigations and identification of safety issues. Further, the FAA has addressed this in TSO-C176a, "Cockpit Image Recorder Equipment," which outlines the minimum performance standard for cockpit image recorder systems based on the EUROCAE Document ED-112A.

Equipment approved under TSO-C176a could accommodate flight data, cockpit voice, and cockpit image recording functions in a single combination recorder. To maximize the potential for recorder survivability and information regarding the cockpit environment, one combination recorder providing these functions would be installed as close to the cockpit as practicable and another as far aft as practicable. The NTSB recommends that the FAA require that all newly manufactured aircraft operated under 14 CFR Part 121 or 135 and required to have a CVR and an FDR also be equipped with a crash-protected cockpit image recording system compliant with TSO-C176a or equivalent. The cockpit image recorder should be equipped with an independent power source consistent with that required for cockpit voice recorders in 14 CFR 25.1457. Additionally, the NTSB recommends that the FAA require that all existing aircraft operated under 14 CFR Part 121 or 135 and currently required to have a CVR and an FDR be retrofitted with a crash-protected cockpit image recording system compliant with TSO-C176a or equivalent. The cockpit image recorder should be equipped with an independent power source consistent with that required for cockpit voice recorders in 14 CFR 25.1457.

Safety Recommendations A-15-5 through -8 recommend similar actions to Safety Recommendations A-00-30 and -31; however, they separately address safeguarding against disabling flight recorders and requiring cockpit image recorders. Issuing TSO-C176a was a good first step; however, after 15 years the FAA still has not mandated installing cockpit image recorders or incorporating protections against the flight crew disabling flight recorder systems. Therefore, Safety Recommendations A-00-30 and -31 are classified "Closed—Unacceptable Action/Superseded."

RECOMMENDATIONS

Therefore, the National Transportation Safety Board makes the following recommendations to the Federal Aviation Administration:

Require that all aircraft used in extended overwater operations and operating under Title 14 *Code of Federal Regulations* (1) Part 121 or (2) Part 135 that are required to have a cockpit voice recorder and a flight data recorder, be equipped with a tamper-resistant method to broadcast to a ground station sufficient information to establish the location where an aircraft terminates flight as the result of an accident within 6 nautical miles of the point of impact. (A-15-1)

Require that all aircraft used in extended overwater operations and operating under Title 14 *Code of Federal Regulations* (1) Part 121 or (2) Part 135 that are required to have a cockpit voice recorder and a flight data recorder, be equipped with an airframe low frequency underwater locating device that will function for at least 90 days and that can be detected by equipment available on military, search and rescue, and salvage assets commonly used to search for and recover wreckage. (A-15-2)

Require that all newly manufactured aircraft used in extended overwater operations and operating under Title 14 *Code of Federal Regulations* (1) Part 121 or (2) Part 135 that are required to have a cockpit voice recorder and a flight data recorder, be equipped with a means to recover, at a minimum, mandatory flight data parameters; the means of recovery should not require underwater retrieval. Data should be captured from a triggering event until the end of the flight and for as long a time period before the triggering event as possible. (A-15-3)

Concurrent with the implementation of Safety Recommendations A-15-1 and A-15-3, coordinate with other international regulatory authorities and the International Civil Aviation Organization to harmonize the implementation of the requirements outlined in Safety Recommendations A-15-1 and A-15-3. (A-15-4)

Identify ways to incorporate adequate protections against disabling flight recorder systems on all existing transport category aircraft. (A-15-5) (Supersedes Safety Recommendation A-00-30)

Require that all newly manufactured transport category aircraft incorporate adequate protections against disabling flight recorder systems. (A-15-6) (Supersedes Safety Recommendation A-00-31)

Require that all existing aircraft operated under Title 14 *Code of Federal Regulations* (CFR) Part 121 or 135 and currently required to have a cockpit voice recorder and a flight data recorder be retrofitted with a crash-protected cockpit image recording system compliant with Technical Standard Order TSO-C176a, "Cockpit Image Recorder

Equipment," TSO-C176a or equivalent. The cockpit image recorder should be equipped with an independent power source consistent with that required for cockpit voice recorders in 14 CFR 25.1457. (A-15-7) (Supersedes Safety Recommendation A-00-30)

Require that all newly manufactured aircraft operated under Title 14 *Code of Federal Regulations* (CFR) Part 121 or 135 and required to have a cockpit voice recorder and a flight data recorder also be equipped with a crash-protected cockpit image recording system compliant with Technical Standard Order TSO-C176a, "Cockpit Image Recorder Equipment," or equivalent. The cockpit image recorder should be equipped with an independent power source consistent with that required for cockpit voice recorders in 14 CFR 25.1457. (A-15-8) (Supersedes Safety Recommendation A-00-31)

The National Transportation Safety Board also classifies the following previously issued recommendations to the Federal Aviation Administration "Closed—Unacceptable Action/Superseded":

Require that all aircraft operated under Title 14 *Code of Federal Regulations* Part 121, 125, or 135 and currently required to be equipped with a cockpit voice recorder (CVR) and digital flight data recorder (DFDR) be retrofitted by January 1, 2005, with a crash-protected cockpit image recording system. The cockpit image recorder system should have a 2-hour recording duration, as a minimum, and be capable of recording, in color, a view of the entire cockpit including each control position and each action (such as display selections or system activations) taken by people in the cockpit. The recording of these video images should be at a frame rate and resolution sufficient for capturing such actions. The cockpit image recorder should be mounted in the aft portion of the aircraft for maximum survivability and should be equipped with an independent auxiliary power supply that automatically engages and provides 10 minutes of operation whenever aircraft power to the cockpit image recorder and associated cockpit camera system ceases, either by normal shutdown or by a loss of power to the bus. The circuit breaker for the cockpit image recorder system, as well as the circuit breakers for the CVR and the DFDR, should not be accessible to the flight crew during flight. (A-00-30) (Superseded by Safety Recommendations A-15-5 and -7)

Require that all aircraft manufactured after January 1, 2003, operated under Title 14 *Code of Federal Regulations* Part 121, 125, or 135 and required to be equipped with a cockpit voice recorder (CVR) and digital flight data recorder (DFDR) also be equipped with two crash-protected cockpit image recording systems. The cockpit image recorder systems should have a 2-hour recording duration, as a minimum, and be capable of recording, in color, a view of the entire cockpit including each control position and each action (such as display selections or system activations) taken by people in the cockpit. The recording of these video images should be at a frame rate and resolution sufficient for capturing such actions. One recorder should be located as close to the cockpit as practicable and the other as far aft as practicable. These recorders should be equipped with independent auxiliary power supplies that automatically engage and provide 10 minutes of operation whenever aircraft power to the cockpit image recorders and associated cockpit camera systems ceases, either by normal shutdown or by a loss of power to the bus. The circuit breaker for the cockpit image recorder systems, as well as the circuit breakers for the CVRs and the DFDRs, should not be accessible to the flight crew during flight. (A-00-31) (Superseded by Safety Recommendations A-15-6 and -8)

Acting Chairman HART, and Members SUMWALT and WEENER concurred in these recommendations.

The NTSB is vitally interested in these recommendations because they are designed to prevent accidents and save lives. We would appreciate receiving a response from you within 90 days, as required by 49 *United States Code* section 1135, detailing the actions you have taken or intend to take to implement them. When replying, please refer to the safety recommendations by number and submit your response electronically to correspondence @ntsb.gov.

[Original Signed]

By: Christopher A. Hart,
Acting Chairman

End Notes

[1] Documents and transcripts from the forum are available in the NTSB Docket Management System at http://dms.ntsb.gov/pubdms/search/hitlist.cfm?docketed=56909&CFID=4303 99&CFTOKEN =77330661 (NTSB Accident ID: DCA14SS009).

[2] See Bureau d'Enquêtes et d'Analyses, *Final Report on the Accident to Air France Airbus 330-203 AF447 from Rio de Janeiro to Paris on 01 June 2009*, (France: Bureau d'Enquêtes et d'Analyses, 2012), which can be accessed at http://www.bea.aero /docspa/2009/f-cp090601.en/pdf/f-cp090601.en.pdf.

[3] Bureau d'Enquêtes et d'Analyses, *Flight Data Recovery Working Group Report,* Technical Document, Issued December 22, 2009, http://www. bea.aero/en/enquetes/flight. af.447/flight. data. recovery.working.group.final.report.pdf.

[4] Office of the Chief Inspector of Air Accidents, Ministry of Transport, Malaysia, *MH 370 Preliminary Report Serial*, 03/2014, http://www.dca.gov.my/MH370/Preliminary%20Report.pdf.

[5] Australian Transport Safety Bureau, *MH370 Progress Report*, http://www.atsb.gov.au/mh370/progressreport.aspx.

[6] Examples of ground stations include ATOs and airline dispatch centers.

[7] As specified in 14 CFR 91.225, all aircraft operating in certain US airspace will be required to carry ADS-B transmitters by January 1, 2020.

[8] *Flight Data Recovery Working Group Report,* http://www.bea.aero/en/enquetes/flight.af.447/flight.data recovery. working. group. final.report.pdf.

[9] Bureau d'Enquêtes et d'Analyses, *Triggered Transmission of Flight Data Working Group Report*, Technical Document, Issued March 18, 2011,http://www.bea.aero/en/enquetes/flight.af.447 / triggered.transmission.of.flight.data.pdf.

[10] "Aircraft Tracking," working paper presented by France at the International Civil Aviation Organization Multidisciplinary Meeting Regarding Global Tracking in Montreal, Canada, May 12-13, 2014. Global Tracking 2014-WP/8 9/5/14, http://www.icao.int/Meetings/GTM/Documents/WP.08.France.Aircraft%20tracking.pdf.

[11] International Civil Aviation Organization, "Conclusions and Recommendations of the Multidisciplinary Meeting On Global Tracking" in Montreal, Canada, May 12-13, 2014, http://www.icao.int/Meetings/GTM/Documents/Final%20Global%20Tracking%20Meeting%20Co nclusions%20and%20%20Recomme ndations.pdf.

[12] "Concept of Operations to Enhance Global Flight Tracking," working paper to be presented by the International Civil Aviation Organization Secretariat at the Second High-Level Safety Conference 2015, "Planning for Global Aviation Safety Improvement" in Montreal, Canada, February 2-5, 2015, HLSC/15-WP/2, http://www.icao.int/Meetings/HLSC2015 /Documents /WP/ wp002 en.pdf.

[13] According to 14 CFR 1.1, an EOW operation occurs over water at a horizontal distance of more than 50 nm from the nearest shoreline.

[14] See 14 CFR 121.339 and 14 CFR 135.167.

[15] The working group participants included representatives from the NTSB, the FAA, foreign government agencies, and industry.

[16] See the Boeing presentation given at the NTSB *Emerging Flight Data and Locator Technology Forum,* October 7, 2014, which is available in the NTSB Docket Management System at http://dms.ntsb.gov/ pubdms/ search/ hitlist.cfm ?docketID=5 6909&CFID =430399& CFTOKEN = 77330661 (NTSB Accident ID: DCA14SS009). The average time for recovery of flight recorders in 27 over water accidents in the past 35 years was about 181 days. The 181-day average does not include the recorders from accidents dating back to 1987 that have yet to be recovered.

[17] Some examples include the FDR from South African Airways Flight 295 (1987) and the CVR and FDR from Asiana Airlines Flight 991 (2011), which were never found. The CVR from Yemenia Airlines Flight 626 (2009) was damaged and some audio was lost. (See *Report of the Board of Inquiry Into the Loss of South African Airways Boeing 747 – 244B Combi Aircraft "Helderberg" in the Indian Ocean on November 28th 1987*; Flight Data Recovery Working Group Report; Flight Data Recovery Working Group Report, http://www.bea.aero/en/enquetes/flight. af.447/flight.data.recovery.working.group.final.report.pdf; Yemenia Airlines Flight 626 report, http://www.bea.aero/docspa/2009/7o-j090629/pdf/7o-j090629.pdf.

[18] See NTSB Safety Recommendation A-99-17 to the FAA, which is classified "Closed—Unacceptable Action." This recommendation letter and excerpts of associated correspondence are available via

the NTSB safety recommendations database at http://www.ntsb.gov/safety/safety-recs/_ layouts/ ntsb.recsearch / RecTabs.aspx.

[19] See the European Organization for Civil Aviation Equipment (EUROCAE) Document ED-112A, *Minimum Operational Performance Specification for Crash Protected Airborne Recorder Systems*, and FAA TSO-123c, "Cockpit Voice Recorder Equipment," and TSO-124c, "Flight Data Recorder Equipment."

[20] The highest data rate FDRs currently available—2,048 12-bit words per second (wps)—would require a data link with a minimum bandwidth of about 25 kbps to stream a complete set of real-time parametric flight data.

[21] The majority of FDRs in use today have data rates of 128 or 256 wps. With these data rates, it is possible to transmit at least 61 seconds of flight data through a bandwidth of 200-400 kbps, which would provide 1 minute of data before a triggering event for every 1 second of data after a triggering event.

[22] Mandatory flight data parameters are specified in 14 CFR 121.343 and 121.344 and 14 CFR 135.152.

[23] See International Civil Aviation Organization, *Operation of Aircraft, Part I, International Commercial, Air Transport—Aeroplanes*, International Standards and Recommended Practices, Annex 6 to the Convention on International Civil Aviation, Ninth Edition (Montreal, Quebec, Canada: International Civil Aviation Organization, July 2010, Amendment 38, November 13, 2014).

[24] National Transportation Safety Board, *In-Flight Fire and Impact With Terrain, Valujet Airlines Flight 592 DC-9-32, N904VJ, Everglades, Near Miami, Florida, May 11, 1996*, AAR-97/06 (Washington, DC: National Transportation Safety Board, 1997), http://www.ntsb.gov/investigations/AccidentReports/Reports/AAR9607.pdf; National Transportation Safety Committee's *Aircraft Accident Report, SilkAir Flight MI 185, Boeing B737-3009VTRF, Musi River, Palembang, Indonesia, December 19, 1997* (Department of Communications, Republic of Indonesia: Jakarta, December 14, 2000) available in the NTSB Docket Management System at http://dms.ntsb.gov/pubdms/search/document.cfm?docID=377242&docketID= 15358&mkey= 11051; Transportation Safety Board of Canada, *Aviation Investigation Report, In-Flight Fire Leading to Collision with Water, Swissair Transport Limited, McDonnell Douglas MD-11 HB-IWF, Peggy's Cove, Nova Scotia 5 nm SW, 2 September 1998*, Report Number A98H0003, http://www.tsb.gc.ca/eng/rapports – reports/aviation/1998/a98h0003/ a98h0003.pdf; National Transportation Safety Board, *EgyptAir Flight 990, Boeing 767-366ER, SU-GAP, 60 Miles South of Nantucket, Massachusetts, October 31, 1999, AAB-02/01* (Washington, DC: National TransportationSafety Board, 2002), http://www.ntsb.gov/investigations/ Accident Reports/Reports/AAB0201.pdf; the NTSB recommendation letter for Safety Recommendations A-00-30 and -31 and excerpts of associated correspondence are available via the NTSB safety recommendations database at http://www.ntsb.gov/safety/safety-recs/_layouts/ntsb. recsearch/ RecTabs.aspx.

[25] National Transportation Safety Board, *Collision with Trees and Crash Short of the Runway, Corporate Airlines Flight 5966, BAE Systems BAE-J3201, N875JX, Kirksville, Missouri, October 19, 2004*, AAR-06/01(Washington, DC: National Transportation Safety Board, 2006), http://www.ntsb.gov/investigations/AccidentReports/Reports/AAR0601.pdf. *Emirates, 03 September 2010*, Case Reference 13/2010.

[26] General Civil Aviation Authority of the United Arab Emirates, Air Accident Investigation Sector Final Air Accident Investigation Report, *Uncontained Cargo Fire Leading to Loss of Control Inflight and Uncontrolled Descent into Terrain, Boeing 747-44AF, N571UP, Dubai, United Arab*

INDEX

A

access, 20, 21, 35, 37, 42, 43, 62, 63
accessibility, 63, 64
advancement, 6
aerospace, 6, 48
Africa, 7
agencies, 5, 40, 41, 49, 69
air carriers, 48
air traffic control system, 48
airline industry, 20, 34
airports, 49
airways, 7
Alaska, 41, 47, 53
appropriations, 51
Appropriations Act, 51
Asia, 7, 52
assessment, 28, 44
assets, 30, 59, 60, 66
ATO, 57
audit, 6
authorities, vii, 1, 2, 4, 9, 15, 17, 28, 29, 30, 40, 49, 52, 57, 62, 66
authority, 4
automation, 51
aviation industry, 6, 19, 41, 42, 44
aviation services, 7

B

Bahrain, 47
bandwidth, 36, 62, 70
batteries, 18, 29, 30, 59
BEA, 4, 9, 17, 29, 30, 35, 52, 56, 57, 58, 61, 64
Beijing, 4, 12, 56
benchmarks, 44
benefits, 20, 21, 22, 27, 28, 34, 39, 40, 52
Brazil, vii, 2, 13, 15, 56
broadband, 61

C

category a, 57, 64, 66
certificate, 49
certification, 10, 52
CFR, 52, 60, 62, 63, 65, 67, 69, 70
challenges, vii, 1, 2, 5, 18, 22, 28, 29, 30, 36, 64
Chicago, 48, 49
China, 12, 17, 47, 56
clarity, 20
clusters, 13
Code of Federal Regulations (CFR), 59, 66, 67

Index

collaboration, 43
collisions, 7, 42
color, 67, 68
commercial, vii, 1, 3, 5, 8, 10, 17, 19, 21, 22, 25, 29, 31, 32, 33, 34, 35, 36, 38, 39, 48, 49, 50, 51, 52, 53, 61
communication, 2, 8, 9, 13, 20, 28, 48, 50
communication systems, 8, 49
communication technologies, 50
community, 2, 18, 19, 39, 43, 49
confidentiality, 64
configuration, 53
Congress, vii, 1, 5, 53
connectivity, 21, 62
consensus, 30
consent, 49
coordination, 2, 3, 9, 13, 15, 17, 19, 20, 23, 25, 28
cost, 2, 3, 5, 18, 19, 21, 22, 23, 27, 30, 34, 37, 39, 51, 52, 56
cost-benefit analysis, 27
country of origin, 32
customers, 52

D

data analysis, 44
data communication, 36
data rates, 70
database, 70
deficiencies, 42
deformation, 61
degradation, 65
delegates, 24
Denmark, 23
Department of Transportation, 6, 48, 49
depth, 2, 18, 31, 59
detection, 18, 59
deviation, 53
disaster, 9, 28
distress, vii, 2, 3, 8, 9, 11, 17, 18, 19, 24, 25, 27, 28, 36, 49
diversity, 34
DOT, 40
draft, 2, 40

DRS, 41, 48, 61

E

emergency, 3, 9, 11, 18, 19, 20, 23, 25, 27, 28, 31, 33, 35, 36, 49, 53, 58
engineering, 22, 34, 52
environment(s), 8, 9, 11, 17, 34, 39, 42, 43, 63, 64, 65
equipment, 2, 10, 12, 18, 22, 27, 30, 33, 34, 37, 45, 48, 51, 59, 60, 63, 66
Europe, 7, 21, 25
European Commission, 58
Everglades, 70
evidence, 6, 64
Executive Order, 52
expertise, 2, 6
exposure, 60

F

FAA, 2, 4, 5, 6, 7, 8, 10, 11, 21, 23, 27, 28, 29, 30, 34, 35, 36, 37, 39, 40, 41, 42, 43, 44, 48, 49, 50, 51, 52, 53, 56, 57, 59, 60, 62, 63, 64, 65, 69, 70
FDR, 4, 9, 10, 29, 31, 32, 35, 36, 37, 38, 39, 52, 56, 59, 60, 61, 62, 64, 65, 69
federal agency, 52
financial, 27
flexibility, 34, 40
flights, 5, 6, 8, 22, 25, 39, 41, 49, 51, 64
fluid, 34
force, 2, 18, 58
France, vii, 2, 4, 15, 46, 48, 56, 57, 59, 64, 68, 69
fuel consumption, 20

G

GAO, 1, 2, 10, 14, 16, 26, 40, 41, 47, 53
global scale, vii, 1
governance, 43, 52
governments, 5, 31
guidance, 7, 10, 11, 28

Index

guidelines, 64

H

Hawaii, 21
hazards, 42
health, 45
history, 5, 6, 9, 39
Hong Kong, 17
House, 53
House of Representatives, 53
human, 34, 37
human remains, 34, 37

I

identification, 42, 57, 58, 59, 62, 64, 65
image(s), 29, 38, 53, 56, 63, 64, 65, 66, 67, 68
imagery, 64
improvements, 44
Indonesia, 46, 47, 70
industry, vii, 1, 2, 5, 18, 23, 24, 27, 29, 31, 33, 34, 39, 41, 42, 44, 49, 55, 57, 58, 62, 63, 69
information sharing, 3, 19, 25, 28
infrastructure, 3, 9, 52
integrity, 37
international standards, 6, 30, 40
Intervals, 11
investment, 51
Ireland, 51
issues, 6, 19, 39, 45, 49, 55, 56, 58, 62, 64, 65
Italy, 23, 47
Ivory Coast, 47

J

Java, 46
Jordan, iii

K

Kenya, 47

L

laws, 49
lead, 15, 39, 42
Lebanon, 46
lithium, 27
localization, 57
location information, 32, 36

M

majority, 70
Malaysia, vii, 2, 4, 16, 17, 46, 48, 50, 56, 69
management, 8, 20, 22, 25, 43, 45, 50, 51, 52
mass, 34
materials, 52
matter, 17
Mediterranean, 46
memory, 34
meridian, 50
messages, 8, 11, 12, 15, 23, 31, 32, 45, 51, 53
Miami, 70
military, 12, 30, 31, 33, 50, 59, 61, 66
Min, Ho Chi, 17
mission, 6, 15, 49, 50
Missouri, 63, 70
misuse, 39
models, 45, 59
modernization, 50

N

NAS, 4, 42, 44, 48
navigation system, 49
next generation, 42
Next Generation Air Transportation System, 4, 21

NextGen, 4, 21, 23, 44, 50, 51
North America, 21

O

oceanic areas, 5, 29, 51
oceans, 11, 23, 48
Office of Management and Budget, 52
officials, 5, 21, 27, 34, 37, 42, 43
OMB, 52
operations, 2, 6, 7, 8, 9, 11, 17, 19, 20, 21, 27, 28, 35, 38, 42, 43, 45, 48, 49, 52, 58, 60, 62, 66

P

Pacific, 47
parallel, 24, 25
participants, 44, 69
passenger airline, 21
permit, 49
Perth, 50
platform, 44
polar, 11
potential benefits, 32, 35, 36
preparedness, 39
prevention, 43
probability, 10
profitability, 41, 45
project, 44, 50
protection, 3, 63

Q

query, 43, 51
questioning, 15

R

radar, 2, 7, 9, 11, 12, 48, 51, 56, 58
radio, 7, 8, 11, 12, 13, 17, 21, 22, 23, 45, 48, 57
radius, 11, 57, 58

real time, 11, 39, 45
reception, 13
recommendations, 5, 18, 20, 24, 38, 40, 53, 55, 56, 57, 63, 64, 66, 67, 68, 70
recovery, vii, 1, 3, 5, 9, 17, 29, 30, 31, 33, 34, 36, 37, 39, 52, 55, 57, 59, 61, 62, 66, 69
redundancy, 60, 61
regulations, 10, 42, 62, 64
regulatory changes, 6
repair, 45
repo, 52
requirement, 58
requirements, 30, 53, 58, 61, 63, 65, 66
resolution, 67, 68
resources, 9, 15, 42, 43, 49, 59
response, vii, 1, 5, 13, 18, 19, 29, 35, 39, 58, 63, 68
risk(s), 27, 32, 33, 42, 43, 44, 49, 59
risk assessment, 43
routes, 7, 21
rules, 28, 49
Russia, 47

S

safety, 3, 5, 6, 21, 22, 27, 31, 33, 34, 35, 38, 39, 41, 42, 43, 44, 49, 51, 53, 55, 56, 62, 63, 64, 65, 68, 70
satellite service, 6, 23, 57
satellite technology, 51
scope, 40
secure communication, 43
security, 36, 38
sensitivity, 5
sensors, 31
service provider, 4, 6, 7, 20, 21, 22, 23, 28, 41, 50, 51, 52
services, 7, 8, 9, 22, 51, 52
shock, 9
shoreline, 69
signals, 9, 33, 59
Singapore, 17
software, 8
solution, 22, 28, 30, 35, 61

South Africa, 69
South America, 7
specifications, 25, 27, 52
speed of light, 48
stakeholders, vii, 2, 3, 5, 6, 19, 20, 21, 22, 23, 24, 27, 28, 31, 32, 33, 34, 35, 36, 37, 38, 39, 41, 44, 45, 51, 62
state(s), 4, 35
structure, 59
surveillance, 2, 3, 7, 8, 11, 21, 23, 24, 39, 48, 51, 52
survival, 59
survivors, 3, 9, 24, 32

T

Taiwan, 46, 47, 52
target, 25, 42, 49
Task Force, 19, 20, 22, 24, 48, 50
teams, 15, 23
technical comments, 2, 40
technological advances, 34
technologies, 2, 4, 8, 18, 19, 20, 21, 22, 28, 31, 39, 51, 55, 58, 62
technology, 2, 5, 7, 23, 25, 28, 29, 31, 32, 34, 40, 50, 57, 58, 60, 61
territory, 49
testing, 3, 10, 29
text messaging, 23
trade, 6, 27, 30, 32, 34, 36, 37, 38, 39, 41
training, 15, 25, 28, 43
trajectory, 18, 19
transcripts, 68
transmission, 3, 5, 8, 18, 19, 22, 25, 30, 31, 33, 35, 36, 37, 57, 61, 62, 69

transport, 6, 21, 22, 23, 36, 41, 49, 50, 51, 57, 64, 66
transportation, 6, 51
triggers, 2, 27, 38
turbulence, 13

U

unit cost, 34, 37
United Airlines, 41
United Nations, 48
United States, v, 1, 6, 9, 27, 37, 49, 51, 53, 58, 68
USA, 46

V

validation, 23, 50, 51
vessels, 56
vision, 44
vulnerability, 44

W

Washington, 48, 49, 52, 53, 55, 70
water, 17, 33, 48, 52, 55, 58, 59, 60, 61, 69
web, 44
West Africa, 46, 47
Western Australia, 50
White Paper, 48
Wi-Fi, 21
worldwide, 19, 28, 50, 51